AF736377

Palgrave Studies in Literature, Science and Medicine

Series Editors
Sharon Ruston
Department of English and Creative Writing
Lancaster University
Lancaster, UK

Alice Jenkins
School of Critical Studies
University of Glasgow
Glasgow, UK

Catherine Belling
Feinberg School of Medicine
Northwestern University
Chicago, IL, USA

Palgrave Studies in Literature, Science and Medicine is an exciting new series that focuses on one of the most vibrant and interdisciplinary areas in literary studies: the intersection of literature, science and medicine. Comprised of academic monographs, essay collections, and Palgrave Pivot books, the series will emphasize a historical approach to its subjects, in conjunction with a range of other theoretical approaches. The series will cover all aspects of this rich and varied field and is open to new and emerging topics as well as established ones.

Editorial board:
Steven Connor, Professor of English, University of Cambridge, UK
Lisa Diedrich, Associate Professor in Women's and Gender Studies, Stony Brook University, USA
Kate Hayles, Professor of English, Duke University, USA
Peter Middleton, Professor of English, University of Southampton, UK
Sally Shuttleworth, Professorial Fellow in English, St Anne's College, University of Oxford, UK
Susan Squier, Professor of Women's Studies and English, Pennsylvania State University, USA
Martin Willis, Professor of English, University of Westminster, UK

More information about this series at
http://www.palgrave.com/gp/series/14613

Joseph Crawford

Inspiration and Insanity in British Poetry

1825–1855

palgrave
macmillan

Joseph Crawford
University of Exeter
Exeter, UK

Palgrave Studies in Literature, Science and Medicine
ISBN 978-3-030-21670-2 ISBN 978-3-030-21671-9 (eBook)
https://doi.org/10.1007/978-3-030-21671-9

© The Editor(s) (if applicable) and The Author(s), under exclusive licence to Springer Nature Switzerland AG 2019
This work is subject to copyright. All rights are solely and exclusively licensed by the Publisher, whether the whole or part of the material is concerned, specifically the rights of translation, reprinting, reuse of illustrations, recitation, broadcasting, reproduction on microfilms or in any other physical way, and transmission or information storage and retrieval, electronic adaptation, computer software, or by similar or dissimilar methodology now known or hereafter developed.
The use of general descriptive names, registered names, trademarks, service marks, etc. in this publication does not imply, even in the absence of a specific statement, that such names are exempt from the relevant protective laws and regulations and therefore free for general use.
The publisher, the authors and the editors are safe to assume that the advice and information in this book are believed to be true and accurate at the date of publication. Neither the publisher nor the authors or the editors give a warranty, express or implied, with respect to the material contained herein or for any errors or omissions that may have been made. The publisher remains neutral with regard to jurisdictional claims in published maps and institutional affiliations.

Cover illustration: Oldtime / Alamy Stock Photo

This Palgrave Macmillan imprint is published by the registered company Springer Nature Switzerland AG
The registered company address is: Gewerbestrasse 11, 6330 Cham, Switzerland

Acknowledgements

Of the four books I have written, this has been much the hardest, mostly due to the fact that it had to compete for attention with a curious, articulate, and highly mobile toddler. My thanks are thus due first and foremost to my wife, Filipa, for endlessly preventing our infant son from tipping laptop computers off desks even when he *really* wanted to, and to my parents, Elaine and Richard, for repeatedly stepping in whenever their help was most required. I also wish to extend my sincere gratitude to Pascale Aebischer, Emily Bernhard-Jackson, Marie Mulvey-Roberts, Andrew Rudd, and Chris Stokes, all of whom read sections of the manuscript and offered invaluable feedback, and to my siblings, Rosa, Oliver, and Sophie, for putting up with me during all those seaside holidays in Devon.

This book is dedicated, with great affection, to my son, Eli, who taught me that even poetic genius would struggle to compete with the effortless imaginative creativity that comes from being three years old.

It's made of walls and glue and water and memories and strongness and leopards!

Contents

CHAPTER 1

Introduction

I first read the Prophetic Books of William Blake as an undergraduate. Having previously known Blake only as the author of 'The Lamb' and 'The Tyger', they came as something of a shock, and I remember feeling mentally bowled over by the enormously authoritative poetic voice that I encountered in works such as *Milton* and *Jerusalem*. When I attempted to share my enthusiasm for Blake with another student, however, his response was damning. After briefly surveying the illuminated books, he concluded that if Blake had been alive today, he would just have been another crazy person writing nonsense manifestos on the internet.

Back then, we undergraduates felt that we knew a thing or two about craziness. Depression, alcoholism, and self-harm were epidemics among us. We knew all about antidepressants and antipsychotics, aspirin overdoses and lithium treatments, counselling sessions and razorblades. We took it for granted that, as sensitive young people studying arts and humanities subjects at university, we would of course be both deeply miserable and slightly mad. In retrospect, I'm afraid we may secretly have been rather proud of it.

I admired Blake because he seemed so sure that he *wasn't* mad. I liked the way that he wrote outrageous things, knowing his readers would accuse him of insanity, and then threw those accusations back in their faces, insisting that *he* was sane and the rest of the world was crazy. His certainty regarding the authenticity of his visions seemed very different from the anxious hesitancy and reflexive irony with which my

© The Author(s) 2019

J. Crawford, *Inspiration and Insanity in British Poetry*,
Palgrave Studies in Literature, Science and Medicine,
https://doi.org/10.1007/978-3-030-21671-9_1

contemporaries and I tended to express our ideas. I tried to explain this to another of my undergraduate friends, but she frowned and replied: 'Are you *sure* he wasn't on drugs?'

In that time and place, such scepticism was predictable. We thought of ourselves as post-modern intellectual sophisticates: doubt was our bread and butter. But we also assumed that our wary, psychiatrically informed perspective was a distinctively modern one. Years later, when I began to study the early poetry of Browning, Tennyson, and their contemporaries, I was thus surprised to find it marked by very similar traits: an admiration of visionary certainty existing side by side with anxious self-doubt, fears of mental instability, and worries over how (or if) one could distinguish between inspiration and insanity. It was one thing for later Victorians, their faith shaken by Strauss and Darwin, to look askance at would-be poet-prophets. But why did the young poets of the 1830s feel such anxieties, with King William on the throne and Shelley barely cold in his grave? Why did they have so much less confidence in poetic intuition, inspiration, and genius than their immediate predecessors? And why did they all seem to have been so worried about going mad?

This book is my attempt to answer these questions. Its subject is the way in which poetic genius was medicalised and pathologised during the second quarter of the nineteenth century, and the impact that this process had upon the development of post-Romantic poetry in Britain, as exemplified by the early works of Browning and Tennyson. From the 1820s onwards, it became increasingly commonplace in Britain for poetic talent to be described in basically medical terms, as a consequence of the unusual physical development of a poet's brain or nervous system—a view which brought with it the related belief that such abnormally poetic brains might also be predisposed to malfunction or madness. Yet this period also witnessed the flourishing of 'Romantic' conceptions of poets as inspired visionaries, and it is the relationship between these two apparently opposed modes of understanding poetic genius that this book aims to explore. How could the poet be simultaneously an inspired prophet *and* a mere case study in atypical neurological development? If poetic talent was associated with both inspired vision and insane hallucination, then how was the would-be poet supposed to tell the two apart?

In the eighteenth century the relationship between poetry and the mental sciences had been much less adversarial, and the interlinked cultures of poetry and psychology that flourished in the 1790s have been ably described in works such as Richardson's *British Romanticism and the*

Sciences of the Mind (2001), Faubert's *Rhyming Reason* (2009), and Ruston's *Creating Romanticism* (2013). I am not the first scholar to note that attitudes towards poetic talent shifted during the post-Romantic period, and in *The Perversity of Poetry: Romantic Ideology and the Popular Male Poet of Genius* (2005), Dino Felluga pointed out that just as poetic ability was persistently linked with melancholy in the eighteenth century, so it came to be connected with madness over the course of the nineteenth century. Felluga linked this change to the ongoing influence of Byron, who was regarded even during his own lifetime as an archetypal 'mad poet' (Felluga 2005, 14–17, 39–41, 108–9, 119–23). The same topic was addressed by James Whitehead in *Madness and the Romantic Poet* (2017), which declared:

> [B]y the turn of the nineteenth century the traditional alignment of genius with the traditional categories of madness, mania, and melancholia, which had descended as protected and even cherished cultural roles from antiquity, was caught up if not entirely overtaken by a more negative medical-moral sense of pathology. (Whitehead 2017, 23–4)

Felluga's and Whitehead's studies confirm the crucial importance of the early nineteenth century, and especially the 1830s, as a transitional moment in the interlinked histories of genius and mental pathology. Both have been important to the present study, and I am particularly indebted to Whitehead's far-ranging analysis of the period. However, my research differs from theirs insofar as my focus is not upon the Romantic poets whose lives and works did so much to bring poetry and insanity together in the British cultural imagination, but upon their immediate successors: the men and women who had to cope with the *consequences* of British Romanticism, which included both the exaltation of poetic genius and the enduring belief that madness and poetry often went hand in hand.

Of these post-Romantic poets I have focused particularly upon the early careers of Tennyson and Browning, whom Isobel Armstrong identified as the chief representatives of the two main schools of early Victorian avant-garde poetry, exemplars of what she called the 'aestheticised politics' of the 'subversive conservative' Apostles and of the 'politicised aesthetics' of the 'dissenting and radical' *Monthly Repository* group, respectively (Armstrong 1996, 27–8). According to Armstrong's classic account, the 1830s saw the rise of the 'Victorian double poem': a form that Browning and Tennyson helped to create, in which poems that first appear to be works of lyric self-expression reveal themselves to be dramatic works that critique the very ideas and ideals they initially seem to express. As Armstrong writes, the double poem was 'a deeply sceptical form' which, by 'seeing utterance both as subject and object', made it 'possible for the poet to explore expressive psychological forms simultaneously as psychological conditions *and* as constructs' (Armstrong 1996, 12–19). My argument is that the rise of such 'double poems' needs to be understood in relation to the period's increasing pathologisation of poetic genius—a development which encouraged poets such as Browning and Tennyson to adopt stances of reflexive self-critique within their poems, thus preempting the possibility that what seemed to be poetic inspiration might turn out to be insane delusion. Within the double poem, whenever the position of subjective lyric speaker threatens to become untenable, the poet can backflip into the stance of an objective observer instead, repositioning themselves as would-be alienists rather than potential patients.

That Victorian poets were interested in psychology is not, of course, a new insight: and two recent studies, Rick Rylance's *Victorian Psychology and British Culture 1850–1880* (2000) and Gregory Tate's *The Poet's Mind* (2012), have greatly deepened our knowledge of the relationship between Victorian psychology and literature. However, Tate and Rylance are both primarily interested in the theoretical psychology which flourished in Britain after 1850, rather than the medical psychology of the previous generation: and while I learned a great deal from both books, I ultimately found them to be of limited assistance in understanding the relationship between poetry and psychiatry in the 1830s and 1840s. The same emphasis upon the post-1850 period, after the great mid-century consolidation of Victorian medical psychology, can be seen in older works on the subject, such as Colley's *Tennyson and Madness* (1983), Faas' *Retreat into the Mind* (1988), and Platizky's *A Blueprint of His Dissent* (1989). Thus Faas argues that 'the dramatic monologue did not establish itself as a proper school of psychological poetry' until 1855, and his study, like those of Colley and Platizky, accordingly hurries towards Tennyson's 'Maud' as quickly as possible, leaving the poetry of the 1830s and 1840s comparatively neglected (Faas 1988, 16). The relationship between poetry and medical psychology in the post-Romantic period has thus tended to fall into the gap between studies on Romantic poetic madness and those on Victorian poetic psychology, and in consequence it has yet to be seriously addressed within the existing scholarship on either subject.

This gap, in turn, means that the question of Tennyson's and Browning's interactions with medical psychology had never been systematically considered in relation to their struggle with the legacies of British Romanticism. This latter topic has been repeatedly addressed in the existing scholarship, in works such as Rivers' *Robert Browning's Theory of the Poet* (1976), Nemoianu's *The Taming of Romanticism* (1984), Tucker's *Tennyson and the Doom of Romanticism* (1988), Harrison's *Victorian Poets and Romantic Poems* (1990), Cronin's *Romantic Victorians* (2002), Baker's *Browning and Wordsworth* (2004), and Martens' *Browning, Victorian Poetics, and the Romantic Legacy* (2011), but none of these works relate the vexed relationship between late Romantic and early Victorian poetry to the changing attitudes towards poetic talent expressed within the medical psychology of the period. However, it is my contention that the shifting interpretation of genius ushered in by the medical psychology of the 1830s and 1840s played an important role in making Nemoianu's 'taming of Romanticism' seem both possible and desirable: and if a 'school of psychological poetry' arose in the 1850s, it was only because the way had been cleared for it by the progressive medicalisation of poetic psychology over the previous generation.

Unlike Faas, Tate, and Rylance, I have thus chosen to focus not on how the mid-nineteenth century 'sciences of the mind' influenced the mature, post-1850 poetry of Browning, Tennyson, and their contemporaries, but on the ways in which their earlier poetry interacted with the medical psychology of the 1825–1850 period. Over the course of these decades, Tennyson and Browning ultimately made their peace with the new psychiatry—but many of their contemporaries did not, and post-Romantic poets such as Barrett Browning, Horne, Townshend, and the 'spasmodics' continued to insist that inspired poetic genius possessed an authority superior to that of any doctor. (Tellingly, many of these writers gravitated towards mesmerism as an alternative to the mainstream medicine of their day.) It is this struggle over the boundaries between inspiration and madness that this book aims to map out.

Methodologically, I have used Tennyson and Browning as case studies of the ways in which the practice of poetry during the period reflected these changing attitudes towards poetic talent. In some ways, the two men were very similar: both were poets of great ability and ambition, both had deep interests in psychology, and both began their careers as admirers of Byron, Shelley, and Keats. In other respects, they were opposites: they moved in very different (and politically opposed) social circles, and their own personal experiences of mental infirmity were poles apart, with Browning seemingly never feeling himself anything less than robustly sane while Tennyson suffered continual anxieties over his own proximity to madness. Their similarities make them comparable, while their differences make the comparisons between them meaningful: for if two poets as socially and psychologically divergent as Browning and Tennyson followed similar creative trajectories over the course of the period, as I shall argue that they did, then this suggests that something more than personal preference was at work. By comparing and contrasting the ways in which their works reflected the evolving debate over inspiration and insanity, I aim to explore how these issues affected individuals for whom the nature of poetic genius was a matter of direct personal significance, rather than an abstract medical or philosophical question. Their importance is heightened by the scale of their influence upon subsequent British poetry—for where Browning and Tennyson went, the next generation of British poets more or less followed. As a result, their responses to the cultural legacies of Romanticism had considerable impact upon the development of Victorian poetry as a whole.

In order to contextualise my readings of Browning and Tennyson, I have examined their works in relation to those of a large number of their contemporaries, ranging from famous writers such as Elizabeth Barrett Browning and John Stuart Mill to lesser-known figures such as Richard Madden, Richard Horne, and Chauncey Hare Townshend. In doing so I have aimed to offer a much broader and more inclusive portrait of the period's attitudes towards the inspiration and insanity of poets than has hitherto been available. Events such as the rise of mesmerism, the vogue for spasmodic poetry, and the creation of the new asylum system are too often considered only in isolation from each other, even though they affected the same people at the same time. My intention here has been to consider them as pieces of a larger picture, thus providing a more adequate account of their relationship with the post-Romantic culture of literary genius as a whole. By considering a relatively narrow chronological period I have been able to discuss a range of controversies which would necessarily have had to be removed from a more 'zoomed-out' study: and if my book consequently lacks the historical breadth of those of Whitehead, Rylance, and Tate, I hope it makes up for it in terms of added depth.

In addition to the introduction and conclusion, this book has five chapters, which are arranged in broadly chronological sequence. Chapter 2 discusses the ways in which the figure of the 'inspired' poet came to be understood in increasingly medicalised terms during the years which followed Byron's death. Chapter 3 explores how Tennyson and Browning responded to the controversies outlined in the previous chapter, and their respective retreats away from their early Shelleyan enthusiasms over the course of the 1830s. The next two chapters discuss slightly later episodes from the lives of these poets as a means of illustrating specific points of cultural tension: Chap. 4 uses the divergent experiences of Tennyson and John Clare in 1837–1842 as a means of exploring the cultural impact of the new asylum system, while Chap. 5 examines Browning's courtship of Elizabeth Barrett in relation to the rising popularity of mesmerism, which Barrett believed in and which Browning regarded with deep scepticism. Finally, Chap. 6 considers the increasingly 'psychological' poetry of Browning's *Men and Women* and Tennyson's *Maud and Other Poems* in relation to the 'spasmodic' poetry of the 1850s, which aimed to reassert the supreme authority of the poet in all matters of poetic inspiration.

Collectively, these chapters chart the rising prominence of a new 'medico-psychological' interpretation of the nature of 'Romantic' poetic genius during this period—a period which also saw the medicalisation and pathologisation of drunkenness, political radicalism, criminality, and fringe religion, all of which had been celebrated by the Romantics themselves. They map out the consequent movement of both Browning and Tennyson from early Romantic faith to deep scepticism regarding the authority of self-declared poetic visionaries. Finally, they explore the ways in which the authors of the period attempted to establish what roles might still exist for poets within this new age of psychological science—even when they were, as in these years they so often seemed to be, beset with what Richard Madden and Isaac D'Israeli tactfully described as 'the infirmities of genius' (Madden 1833, title; D'Israeli 1822, x).

References

Armstrong, Isobel. 1996. *Victorian Poetry*. London: Routledge.

D'Israeli, Isaac. 1822. *The Literary Character, Illustrated by the History of Men of Genius*. 3rd ed., 2 vols. London: John Murray.

Faas, Ekbert. 1988. *Retreat into the Mind: Victorian Poetry and the Rise of Psychiatry*. Princeton: Princeton University Press.

Felluga, Dino. 2005. *The Perversity of Poetry: Romantic Ideology and the Popular Male Poet of Genius*. New York: New York State University Press.

Madden, Richard. 1833. *The Infirmities of Genius*. 2 vols. Philadelphia: Carey, Lee, and Blanchard.

Whitehead, James. 2017. *Madness and the Romantic Poet: A Critical History*. Oxford: Oxford University Press.

CHAPTER 2

'He was not one of ye': Poetry and Mental Peculiarity, 1825–1836

This chapter explores how the conceptions of poetic genius fostered by the writings of Byron and Shelley served to reinforce existing stereotypes of poets as antisocial and mentally abnormal. This belief, coupled with the increasing confidence of contemporary physicians that all mental traits could be explained in physical terms, led to the widespread assumption that poetic talent was due to the abnormal development of the brain and nervous system. In consequence, the belief arose that poets were naturally predisposed to suffer from mental and physical illness, and that great poetic talent would naturally be accompanied by great vulnerability to insanity. This complex of linked ideas can be seen in the works of writers as varied as Isaac D'Israeli, John Stuart Mill, Arthur Hallam, Richard Horne, Richard Madden, and John Wilson Croker, and formed an important part of the intellectual context within which Browning and Tennyson began their respective poetic careers.

The Invention of the 'Romantic Genius'

The period from 1825 and 1836, in which the early works of 'Victorian' writers such as Barrett, Tennyson, and Dickens appeared alongside the last publications of 'Romantic' writers such as Coleridge, Lamb, and Hazlitt, occupies an awkward place in British literary history. No longer properly Romantic and not yet truly Victorian, it exists as an indeterminate borderland between the two: what North calls 'a no-man's-land between epochs',

© The Author(s) 2019
J. Crawford, *Inspiration and Insanity in British Poetry*,
Palgrave Studies in Literature, Science and Medicine,
https://doi.org/10.1007/978-3-030-21671-9_2

and which Cronin describes as 'a shadowy stretch of time sandwiched between two far more colourful periods' (North 2009, 6; Cronin 2002, 1). As North and Higgins have made clear, however, it was precisely during these 'shadowy' years that the idea of 'Romantic' poetic genius was consolidated (North 2009, 5; Higgins 2005, 10). As Higgins writes:

> By the 1830s, a gap had developed between an emergent model of the professional man of letters—meant to be a productive member of civil society—and an account of literary genius—focused mainly on poets—which emphasised the inevitable separation, perhaps even antagonism, between great writers and normal people. (Higgins 2005, 10)

Higgins argues that this new conception of genius was largely disseminated through the periodical culture of the period: as Whitehead notes, the figure of the 'mad poet' was constructed in self-conscious opposition to the figure of the sane reviewer (Whitehead 2017, 12–13). North, meanwhile, has demonstrated the important role played by poetic biographies in popularising the idea that poetic geniuses tended to be very peculiar people. Noting the dominance of the '*Life and Works*' format among editions of poetry published during the 1830s, he argues that 'In the biographical edition a competitive interdependency of poetry and biography became inscribed in the means of disseminating the idea of the "Romantic poet"' (North 2009, 48).

Of these poets, Byron was by far the most famous. Browning and Tennyson were 12 and 14, respectively, at the time of Byron's death, and his poetry made a powerful impression on them as adolescents (Kennedy and Hair 2007, 18; Martens 2011, 24–5; Ricks 1989, 12–13). It was from Byron that they first derived the idea that the ideal poet should be a man unlike other men: a 'wandering outlaw of his own dark mind' like Manfred or Childe Harold, heroic, gloomy, and sublime (Byron 1921, 203). Byron's influence is visible throughout most of the juvenilia written by Tennyson and his brothers during the 1820s, suffusing such poems as 'I Wander in Darkness and Sorrow' and 'The Vale of Bones', in which lonely, darkly brooding speakers describe their wanderings across landscapes whose desolation mirrors their own mental and spiritual states (Tennyson 1987, 1:104–5, 108–11). Many years later, Tennyson regretfully remarked: 'As a boy I was an enormous admirer of Byron, so much so that I got a surfeit of him, and now I cannot read him as I would like to do' (Hallam Tennyson 1897, 2:69). A direct comparison with Browning is impossible, as he burned his earliest poetic works in 1884, but his subsequent remarks about his adolescent poems suggest that they, too, were strongly Byronic in character (Kennedy and Hair 2007, 18; Thomas 1982, 22).

As Felluga and North both emphasise, Byron's life and legend did a huge amount to popularise the idea that great poets were naturally predisposed towards melancholy and eccentricity (Felluga 2005, 108–9, 119; North 2009, 61, 72–5). As early as 1813, the minor poet Robert Gillies wrote an imitation of Byron called *Childe Alarique, A Poet's Reverie*, with the aim of demonstrating that every 'truly poetic mind' was subject to 'despondency', by which 'Some of the noblest intellects have been overthrown' (Gillies 1814, 73–4). North notes that Byron's early biographers presented his genius as both an explanation of and an excuse for his unusual and antisocial behaviour: thus Moore wrote of the 'unsocial and detaching tendency' of poetic greatness, while Gordon asserted that Byron's genius exempted him from normal standards of judgement, insisting that 'the same measure which applies to a dull clerk or a posing lecturer, should not be applied to such a man as him' (North 2009, 73, 75). These attitudes were both endorsed and satirised by Benjamin Disraeli's 1837 novel, *Venetia*, whose protagonist, Lord Carducis, self-consciously sets himself up as a 'sublime' Byronic poet, so different to other men that he is 'Alone in a crowd'. Lamenting Carducis' radical opinions, one minor character remarks that 'He is almost a rebel! But then he is such a genius! Everything is allowed, you know, to a genius!', while another declares:

> He must be a man of genius; he is so unlike everybody; the very tie of his cravat proves it. And his hair, so savage and dishevelled; none but a man of genius would not wear powder. Watch him to-day, and you will observe that he will not condescend to perform the slightest act like an ordinary mortal. (Disraeli 1870, 74, 79)

By the time Carlyle wrote *Sartor Resartus* in 1833–1834, he mockingly assumed that his 'less philosophical readers' would view madness, suicide, and the writing of 'Satanic' (i.e. Byronic) poetry as virtually interchangeable expressions of a passionate soul suffering under worldly disappointments, all three of which were likely to be attended by the same alarming forms of antisocial behaviour:

> To our less philosophical readers, for example, it is now clear that the so passionate Teufelsdröckh, precipitated through 'a shivered Universe' in this

> extraordinary way, has only one of three things which he can next do: Establish himself in Bedlam; begin writing Satanic Poetry; or blow-out his brains. In the progress towards any of which consummations, do not such readers anticipate extravagance enough; breast-beating, brow-beating (against walls), lion-bellowings of blasphemy and the like, stampings, smitings, breakages of furniture, if not arson itself? (Carlyle 1975, 112–3)

In 1820s Britain, there was nothing unusual about being an admirer of Byron, as the low prices at which his poems were sold ensured they were widely read (Felluga 2005, 90–1). Unlike many of their contemporaries, however, Browning and Tennyson also read Keats and Shelley, at a time when the poetry of both men was very little known. Browning's introduction to Shelley came at the age of 15, when his cousin bought him a pirated edition of the *Miscellaneous Poems*: he was so impressed by them that he prevailed upon his mother to buy all the other published works of Shelley, scarce though many of them were at the time, and on her bookseller's suggestion she also bought him the poems of Keats (Kennedy and Hair 2007, 19–20; Martens 2011, 25; Baker 2004, 23; Keenan 1973, 120–1). Tennyson seems to have encountered Keats and Shelley at a similar age, as several echoes from their works appear in his early poem 'Armageddon' (Tennyson 1987, 1:74–85). However, his serious engagement with their poetry began during his time at Cambridge, where he befriended Arthur Hallam and joined the Cambridge Apostles (Ricks 1989, 28; Pearsall 2008, 58). The Apostles were great admirers of Wordsworth, Keats, and Shelley, and the society played an important role in publicising the works of all three poets. In 1829 the Apostles arranged for Shelley's *Adonaïs* to be published for the first time in England, and in a debate with the Oxford Union in the same year they maintained that Shelley was an even greater poet than the then universally lauded Byron—a line of argument which confused their Oxford opponents, most of whom were completely unfamiliar with Shelley's poetry (Cronin 2002, 149–51; Day 2005, 10; Lourie 1979, 5; Higgins 2005, 76–7). Tennyson's admiration of Shelley reached new heights during his time at Cambridge, and he soon displaced Byron as the poet whose example Tennyson most aspired to follow.

For both Browning and Tennyson, the impact of Shelley on their late teens was just as powerful as that of Byron had been a few years before. Decades later, when his enthusiasm had cooled, Tennyson remarked: 'Nobody admires Shelley more than I once did' (Hallam Tennyson 1897,

2:70). Browning found Shelley's arguments on religion and morality so persuasive that he became both an atheist and a vegetarian, although neither of these convictions lasted more than a couple of years (Thomas 1982, 23–4). Meanwhile Hallam and Tennyson, inspired by Shelley's political ideals, set off in 1830 to join the revolutionary struggle against King Ferdinand of Spain. Their involvement in the revolution was brief and minor, but their fellow Apostle Robert Boyd joined the rebel army of General Torrijos, and was ultimately executed for his role in the rebellion in 1831—an event which Armstrong has called 'one of the unspoken repressed elements of Tennyson's early poetry' (Ricks 1989, 52–3; Cronin 2002, 152–4; Armstrong 1996, 31).

In Shelley's view, being a true poet often involved setting oneself apart from the mass of humanity—for, as he made clear in *Adonaïs*, such poets had to be willing to follow their inspiration wherever it led, no matter how far it carried them from the 'trembling throng' of ordinary mortals (Shelley 1977, 406). This ideal of the poet as a mighty mental traveller clearly held enormous appeal for the young Tennyson and Browning. Tennyson's 1830 poem 'The Mystic' gives a portrait of the type of poet that Shelley's example had made him aspire to be:

> Angels have talked with him, and showed him thrones:
> Ye knew him not: he was not one of ye,
> Ye scorned him with an undiscerning scorn:
> Ye could not read the marvel in his eye [...]. (Tennyson 1987, 1:252)

Another example is provided by Browning's famous address to Shelley's spirit in *Pauline*, written in 1832:

> Sun-treader—life and light be thine for ever;
> Thou art gone from us—years go by—and spring
> Gladdens, and the young earth is beautiful,
> Yet thy songs come not—other bards arise,
> But none like thee—they stand—thy majesties,
> Like mighty works which tell some Spirit there
> Hath sat regardless of neglect and scorn,
> Till, its long task completed, it hath risen
> And left us, never to return [...]. (Browning 1983–2009, 1:34)

This Shelleyan cult of the quasi-divine poet exaggerated the tendencies which both men had already absorbed from Byron. If reading Byron had persuaded them that the ideal poet should not be a man like other men, then reading Shelley led them to conclude that he should barely be a man at all: he should be 'some Spirit', one whom 'Angels have talked with', his eyes fixed firmly upon Eternity. (I use the male pronoun deliberately: the idea that such a poet might be a woman did not yet seem to have occurred to either of them.) He will probably be doomed to neglect within his own lifetime, because he will have raised himself so far above the rest of humanity that most people will be unable to understand him: both Browning and Tennyson use the same word, 'scorn', to describe how such poets are received by the 'undiscerning' public, whose hopelessly ordinary minds are unequal to the task of understanding the works of such a man. If Byron's poet-heroes are not like other men, Browning's 'sun-treader' Shelley is not even like other *poets*: as he writes, 'other bards arise, / But none like thee'. He possesses a capacity for visionary insight which sets him apart from, and above, all other men and women, marking him out as 'not one of ye' and making him 'regardless' of ordinary human affairs.

The early 1830s, then, saw the spread of 'Romantic' conceptions of the poet as psychologically different from, and superior to, all other people—conceptions which both Browning and Tennyson enthusiastically embraced. However, these same years also witnessed the collision of this ideal with a newly medicalised approach to the study of the human mind, whose practitioners were growing rapidly in confidence and ambition at the time. This movement, whose approach was typified by the writings of up-and-coming mad-doctors such as Browne, Conolly, and Morison, increasingly asserted that all forms of unusual mental activity could and should be understood in medical terms, and poetic genius was to prove no exception to this general rule.

Genius Explained: Poetry and Medicine in the Post-Romantic Period

As Richardson has shown in *British Romanticism and the Science of the Mind* (2001), the Romantics actively engaged with the mental sciences of their own day. However, while Wordsworth, Coleridge, and Keats were all intrigued by the possibilities opened up by the scientific study of the nerves and brain, Whitehead notes that they generally resisted the idea that poetry and poetic genius could be 'explained' in medical terms, and especially the idea that it might constitute a form of madness (Whitehead

2017, 50). Whitehead observes that Coleridge distinguished between the degenerate madness of 'the Insania pseudo-poetica' and the 'Regenerate' madness of the true 'Furor divinus', maintaining that while the former might be medically accounted for, the latter transcended such reductive explanations (Whitehead 2017, 31). Similarly, Richardson points out that it was precisely Coleridge's recoil from the idea that artistic creativity could be reduced to the simple interplay of organs within the brain which turned him against the phrenological theories of Spurzheim and Gall, for whom *Dichtergeist* ('poetic talent') was merely one of the brain's 27 organs. Coleridge insisted that true poetry 'brings the whole soul of man into activity', and that it thus required the use of every human faculty, not just the freakish over-development of a single region of the brain (Richardson 2001, 23, 48–9).

Ruston and Allard have both noted that Wordsworth saw the poet as 'a man who is different physiologically, intellectually, and emotionally from other men' (Ruston 2013, 24; Allard 2007, 46). However, when Wordsworth defined the qualities of true poets in his prefaces to *Lyrical Ballads*, he did so in almost entirely non-medical terms, emphasising that what poets required was not a specific sort of nervous system, but 'a greater knowledge of human nature, and a more comprehensive soul, than are supposed to be common among mankind'. '[F]rom practice', Wordsworth wrote, '[the poet] has acquired a greater readiness and power in expressing what he thinks and feels, and especially those thoughts and feelings which, by his own choice, or from the structure of his own mind, arise in him without immediate external excitement' (Wordsworth 1992, 751). But it does not follow from this that it is 'the structure of his own mind' which makes him a poet, still less that this 'structure' is due to the actual physical architecture of his brain. Instead, it is the way in which the poet has learned to *express* the 'thoughts and feelings' generated by his mind's structure which constitute his poetic talents. Even if the 'structure of his own mind' is a relatively ordinary one, he can still be a great poet if he possesses a sufficiently impressive 'power in expressing what he thinks and feels'.

Of all the major British Romantics, Shelley came closest to offering a medicalised model of poetic talent. Ruston has described his belief that poets were separated from 'the herd of mankind' by their refined and delicate sensibility, which also made them unusually susceptible to illness and pain: but 'sensibility', in the Romantic-era sense of the word, was far from being a purely physiological quality (Ruston 2005, 132, 135). In his

Defence of Poetry, Shelley asserted that the true poet 'is more delicately organized than other men, and sensible to pain and pleasure, both his own and that of others, in a degree unknown to them'. 'Organization', in this context, could refer to the construction of the poet's body: but given that elsewhere in the *Defence* Shelley writes that poets are 'spirits of the most refined organization', he was probably referring primarily to their *spiritual* 'organization', rather than to the way in which their nervous systems happened to be assembled (Shelley 1977, 505, 507).

From the 1820s onwards, however, two interconnected ideas began to be expressed in Britain with increasing frequency. The first idea was that poetic genius could, as Gall had argued in the 1790s, be understood in physiological terms. Godwin, who in 1797 had believed that poetry could be learned like any other skill, had by 1831 been led by the neurological theories of the period to conclude that poetic talent was in fact the result of an innate physiological predisposition (Ruston 2013, 101–2). Faubert has shown that the rise of phrenology strongly contributed to the belief that the creativity of poets could be explained in terms of their atypical brain development: in her view, it was for this reason that nineteenth-century psychologists stopped writing poems about psychology and started writing treatises about the psychology of poets, having shifted from seeing poetry as a means by which to explain medicine, to seeing poetry as something which required medical explanation (Faubert 2009, 27–8, 196). Kurshan notes that British phrenologists attempted to demonstrate, by measuring the skulls of poets, that 'what is called inspiration' was in fact a consequence of their unusually large organs of 'ideality', and in 1831 the phrenologist Robert Cox even dug up the skull of Burns in order to prove that 'Burns' poetic genius could be attributed to his extraordinary cerebral developments' (Kurshan 2006, 30–1).

The second idea was that poetic talent, by its very nature, made those who possessed it exceptionally vulnerable to insanity. This belief can be seen in Macaulay's 1825 assertion that writing poetry implied and required 'a certain unsoundness of mind', Madden's 1833 writings on 'the morbid sensibility of genius', and Croker's claim in the same year that the brain of a poetic genius 'may be more liable to be deranged by sudden shocks, or impaired by over use, than less delicate organs', to the point where it could even be said that 'such genius is of itself a disease' (Macaulay 1843, 7; Madden 1833, 1:17; Croker 1834, 51). The rise of these two ideas resulted, as Faas notes, in a period in which '[alienists] told poets and critics that genius is, at least in principle, analogous to idiocy and madness; or that the poetic imagination is essentially characterised by abnormality and "organic flimsiness"' (Faas 1988, 9).

As Wood writes, in the early nineteenth century the British middle classes shifted from seeing 'sensitivity and delicacy' as 'marks of civilized refinement' to regarding them as 'sign[s] of latent mental disease' (Wood 2001, 81). Writing in 1829 to the child prodigy Mary Browne, Letitia Landon warned her that 'Genius, like all heavenly light / Can blast as well as bless, the sight', and expressed the hope that, in her case, poetic talent would not be accompanied by its 'usual curse' of madness, sickness, and early death: 'And may for once the laurel wreath / Not wither all that grows beneath!' (Landon 1829, 289, 294). Landon herself was described in 1839 by Emma Roberts as someone whose 'mental sufferings' had been intensified by 'the fine susceptibility of her nervous system'—the same 'fine susceptibility' which had presumably made her such an exceptional poet—while in her preface to the 1840 edition of Shelley's essays, Mary Shelley explained that her husband had never finished his 'complete theory of mind' because 'His nerves, indeed, were so susceptible, that these intense meditations on his own nature, thrilled him with pain' (Landon, *Zenana*, 1839, 14; Shelley 1840, xvii). Unable to complete his own theory of psychology because of the abnormal sensitivity of his nervous system, Shelley exemplifies the way in which, as Faubert has argued, poets came during this period to be seen 'as the subjects, rather than the masters, of psychological discourse' (Faubert 2009, 196).

The increasing popularity of these interconnected ideas arose partly from the phrenologically influenced psychiatry of the period, which insisted that, as Morison trenchantly asserted in 1828, 'In every case of mental derangement, it is presumed that more or less corporeal disorder exists' (Morison 1828, 4). Foucault notes that, from 1825 onwards, the institution of regular autopsies in asylums implied a belief that the 'truth of madness' resided not in the words of the mad, but 'in their nerves and their brain', while Tate argues that 'mental processes increasingly came to be understood and discussed in bodily terms', with the result that 'madness came to be seen as an organic disease which, like physical illnesses, fell under the remit of the medical profession' (Foucault, *Power*, 2006, 249; Tate 2012, 7, 114). If poets were, as tradition claimed, unusually vulnerable to madness, then this implied that something was at fault in the organisation of the poet's body, either due to the strains that the practice of poetry placed upon their nerves, as Madden claimed, or because of the unusual development of their phrenological organs, as Croker suggested.

But the willingness of contemporary writers to believe that poets were very often madmen was also a reflection of the way in which madness itself was progressively redefined over the course of the early nineteenth century.

In the eighteenth century, the lunatic had been defined, by law and medical practice alike, as a *furiosus*, an individual completely incapable of engaging in the normal business of life (Ingram 1992, 28; Prins 1995, 12; Jacobs 1971, 27). When judged by such standards, transgressive writers such as Byron were more likely to be viewed as odd and sinful rather than actually mad. However, the changing moral standards and developing scientific discourses of the nineteenth century led many forms of mental activity which had previously been viewed as relatively ordinary to be reclassified as pathologies requiring legal or medical intervention. Between 1800 and 1845, the boundary lines between sanity and madness were repeatedly redrawn, steadily widening the grounds upon which a person might be deemed mad rather than simply wicked or strange (Whitehead 2017, 75). In consequence, the number of people classed as lunatics rose sharply, and by the 1840s there was a widespread sense of anxiety that British society was at risk of being swamped by a rising tide of insanity (Porter 1987, 116; Shuttleworth 1996, 34, 46–7).

Works on medical psychology written between 1815 and 1830, such as Reid's *Essays on Insanity* (1816), Morison's *Outlines of Lectures on Mental Diseases* (1826) and *Cases of Mental Disease* (1828), Man Burrows' *Commentaries* (1828), and Conolly's *Indications of Insanity* (1830), testify to this newly expansive sense of what might constitute 'mental disease'. Reid wrote that 'Almost every nervous affection may be considered as an approach to insanity': he warned his readers that their minds might contain 'minute embryos of insanity which cannot be discerned by the naked or uneducated eye', especially if they happened to be 'endowed with an exquisite sensibility' (Reid 1817, 162–3). Man Burrows argued that unchecked eccentricity could 'ripen into perfect insanity', especially among geniuses, as 'Genius, improperly directed, is more prone to aberration than an understanding of mediocrity' (Man Burrows 1828, 44, 101–3). It was for similar reasons that Conolly, one of the leading advocates of the new medical psychology, took a very suspicious view of eccentric behaviour:

> Eccentricity in fact is connected with a disease of the comparing faculty, and any affliction of that faculty brings a man nearer to the condition of a lunatic. It is often associated with diseased sensations, and we shall see that

> Insanity consists of a diseased state of one or more of the faculties of mind, involving a loss of the power of comparing. A very eccentric man, therefore, is always a near neighbour to the madman. (Conolly 1830, 135–6)

Conolly's decision to devote an entire chapter of his book to 'Inequalities, weaknesses, and peculiarities of the human understanding, which do not amount to insanity' testified to his sense that sanity and madness, rather than being starkly divided from one another, in fact shaded into one another across a vast and debatable borderland (Conolly 1830, 93). Indeed, the line that divided them was often so very unclear that expert judgement was needed to locate it. In the eighteenth century, ordinary juries had been assumed to be capable of judging whether someone was mad or sane: but in 1830 Conolly wrote that 'all the erroneous views of the nature of insanity, and all the most serious of the errors which have been committed in the management of the insane, have had inattention to these distinctions [between conditions which do and do not "amount to insanity"], or ignorance of them, for their source', strongly implying that the 'distinctions' between madness and sanity were anything but self-evident (Conolly 1830, 93–4; Ingram 1992, 91–3; Taylor 1988, 45). As Shuttleworth notes, 'Nineteenth-century physiology and psychiatry broke down earlier absolute divides between the normal and pathological' (Shuttleworth 1996, 12, 15). Judged by these standards, it became much easier to view the unusual behaviour of poets as indicators of their madness—and thus as demonstrations that there must be something abnormal about the construction of poetic brains.

A comparison of the 1794 first edition of Isaac D'Israeli's *Essay on the Literary Character* to the expanded third edition of 1822 demonstrates how much stronger the interlinked beliefs in the physiological origins of poetic talent and the proximity of poetic genius to madness had become over the intervening years. In 1794 D'Israeli had seen the 'literary character' as simply a personality type, but by 1822 he had developed a far grander sense of the qualities which set literary 'men of genius' apart from ordinary people. The essay's third edition made much greater claims for such individuals, asserting that their 'occupations, amusements [and] ardour' are 'at discord with the artificial habits of life', where 'crowds of human beings are only treading in one another's steps' (D'Israeli 1822, 1:132). D'Israeli now maintained, as he had not in 1794, that a true literary genius would be an enthusiast and visionary, whose 'senses are excited by a mixture of reality with imagination', and that it was in this 'equivocal state' of semi-hallucination that 'the enthusiast of genius produces his masterpieces' (D'Israeli 1822, 2:2, 2:5–6).

If D'Israeli's new edition placed much greater emphasis on the specialness of such men of genius, however, it also had much more to say about the 'disorders' that such literary characters were prone to, leading D'Israeli to wonder whether 'I have sometimes apologised too warmly for the infirmities of men of genius' (D'Israeli 1822, ix–x). Adding new chapters on 'the irritability of genius' and 'the enthusiasm of genius', D'Israeli described their reveries, trances, hallucinations, self-destructive activities, and 'morbid sensibilities', which were 'often weak to effeminacy, and capricious to childishness' (D'Israeli 1822, 1:138–9, 1:278–80, 2:32–6). Such literary men, he wrote, were vulnerable to illness and madness because their mental labours wore out their minds and were incompatible with healthy exercise. He disagreed with Beattie's judgement that being 'too much troubled with delicate nerves' was not a mark of genius, and described the creative passion of such an individual as 'a fever [that] shakes his spirit, a fever which has sometimes generated a disease, and has even produced a slight perturbation of the faculties' (D'Israeli 1822, 1:62–3, 1:134).

The explanation for all this strange and alarming behaviour, D'Israeli claimed, was to be found in the unusual nerves and brains of literary men. In a passage added to the 1822 edition, he wrote:

> Such minds identified themselves with their visions! If we pass them over by asserting that they were insane, we are only cutting the knot which we cannot untie; we have no right to deny what some tell us, that a sympathy and perception of the corporeal with the incorporeal nature of man, his imaginative with his physical existence, is an excitement which appears to have been experienced by persons of a peculiar organization. (D'Israeli 1822, 2:29)

That the 'peculiar organization' which enabled such visionary perception of the 'incorporeal' was a physical trait is made clear by D'Israeli's subsequent assertion that 'The history of great visionaries, were they correctly detailed, would probably prove how their delusions consisted of the ocular spectra of their brain, and the accelerated sensations of their nerves' (D'Israeli 1822, 2:31). Pursuing this idea further, D'Israeli came very close to arguing that literary genius was simply a very specific form of brain disease, writing that '[Kotzebue] believed, what a physiological history would perhaps show, that there are some maladies, those of the brain and the nerves, which actually stretch the powers of the mind beyond their usual reach' (D'Israeli 1822, 32). D'Israeli's tentative language here—his references to what a 'correctly [i.e. medically] detailed' history of genius would '*probably* prove', and to what 'a physiological history would *perhaps* show' (my italics)—reflects his awareness that, in 1822, the medical investigation of creative genius in terms of brain structure and mental disease was still in its infancy. Over the next two decades, however, such attempts to explain artistic creativity in medical terms would become increasingly common in Britain.

The belief in the basically neurological character of poetic talent often existed alongside the most exalted beliefs in the nature of poetic genius. Arthur Hallam had the very highest opinions of Keats and Shelley as poets; yet Tate notes that, in 1830, his reading of Hartley led him to conclude that 'poetry originates in the poet's physical organs as they tremble into motion and emotion' (Tate 2012, 23). In his 1831 essay 'On Some of the Characteristics of Modern Poetry', Hallam asserted that the talents of Keats and Shelley had been due to the unusual nature of their bodies:

> Susceptible of the slightest impulse from external nature, their fine organs trembled into emotion at colours, and sounds, and movements, unperceived or unregarded by duller temperaments. Rich and clear were their perceptions of visible forms; full and deep their feelings of music. So vivid was the delight attending the simple exertions of eye and ear, that it became mingled more and more with their trains of active thought, and tended to absorb their whole being into the energy of sense. Other poets seek for images to illustrate their conceptions; these men had no need to seek; they lived in a world of images; for the most important and extensive portion of their life consisted in those emotions which are immediately conversant with the sensation. (Hallam 1862, 296)

Hallam thus attributed the poetic brilliance of Keats and Shelley to the superiority of their 'fine organs', which seems here to refer primarily to their eyes, ears, and brains, and describes them as having been 'men of very superior make' who were 'born poets' (Hallam 1862, 295). Outfitted with such superior physical equipment, they were able to continuously and effortlessly bridge the gap between sensation and thought, which less fortunate poets could cross only with considerable labour. Such poetic gifts were a 'rare and exclusive privilege', and 'the heights and depths of art are most within the reach of those who have received from nature the "fearful and wonderful" constitution we have described' (Hallam 1862, 297–8).

Like Tennyson and Browning, Hallam saw such 'born poets' as being almost supernatural: their poetry, he wrote, was 'a sort of magic' (Hallam 1862, 298). For Hallam, the 'poet of sensation'—which he clearly viewed as the highest and truest form of poet—was thus a kind of marvellous physiological mutant, a freak of nature lucky enough to be born with physical attributes that allowed him or her to reach levels of poetic attainment that most other writers could never hope to achieve. Small wonder, then, that the minds of such poets should be very different to those of ordinary men and women—so different, in fact, that most people would not be able to understand them at all, leading Hallam to conclude that poets would usually be alienated from society, because 'the public very naturally derided them as visionaries' (Armstrong 1996, 171; Hallam 1862, 298). The very superiority of a poet's 'organs' and 'constitution' doomed them to be lonely and misunderstood. As Hallam ruefully reflected:

> How should they be popular, whose senses told them a richer and ampler tale than most men could understand, and who constantly expressed, because they constantly felt, sentiments of exquisite pleasure or pain, which most men were not permitted to experience? (Hallam 1862, 298)

The View from the Fox Circle: Richard Horne and John Stuart Mill

Robert Browning had no direct contact with Hallam and the Apostles: as a Dissenter, he was barred from studying at Oxford or Cambridge. As Armstrong has emphasised, his literary and intellectual development took place within a very different context: the circle of politically liberal writers and thinkers who gathered around the Unitarian minister and writer William Fox, of whose journal, *The Monthly Repository*, the young Browning was a regular reader (Armstrong 1996, 28; Cronin 2002, 165). It was Fox who persuaded Browning not to try to publish the adolescent poems that he had written in imitation of Byron, but this act—which Browning, in later years, concluded had probably been for the best—was by no means a sign that what would now be called 'Romantic' poetry was not held in high regard among his circle. Fox was a close friend of the radical printer and journalist Benjamin Flower, who served as a living connection with the earliest and most radical years of the Romantic movement, having published Coleridge and Southey's revolutionary tragedy

The Fall of Robespierre in 1794. When Flower died in 1829, Fox became the guardian of his two daughters, Sarah and Eliza, the latter of whom Browning seems to have been in love with throughout most of his adolescence. As well as Flower, the Fox circle included Thomas Talfourd, who was a friend of Lamb and Coleridge; Richard Horne, the Keats-loving poet and critic who succeeded Fox as editor and proprietor of the *Monthly Repository* in 1836; and John Stuart Mill, who had become a deep admirer of Wordsworth's poetry after it helped him to recover from his nervous breakdown in 1826 (Kennedy and Hair 2007, 18–20, 42, 47–8, 52; Gill 1998, 44–51).

Of these men, Horne was much the most outspoken in his praise for what he saw as true poetic genius. In 1833, having failed to find a publisher for his own literary works, he wrote a book-length *Exposition of the False Medium and Barriers Excluding Men of Genius from the Public*, in which he grandly divided the human race into two categories: men of genius (of whom he clearly considered himself to be one) and everyone else. According to Horne, 'the present march of intellect' was 'solely to be attributed to men of genius and ability', and yet such 'men of genius' were universally neglected and misunderstood, due to the fact that 'such a man must always seem eccentric in society, because he is more natural and original than the rest, who, in the average, are only second-hand copies of artificial forms' (Horne 1833, 4, 121–2). Citing Keats as an example of such neglected genius, Horne argued that Shelley, like Keats, would have died a martyr to literature had he not been independently wealthy, and asserted that even Byron might have suffered a similar fate had he lacked the advantages conferred by his aristocratic status (Horne 1833, 27–8, 128).

Like Hallam, Horne saw genius as being an innate, unlearnable, and basically physiological quality, arising from 'that peculiar nervous system which so fatally appertains to intellectual men'—a nervous peculiarity which, in his view, also made them vulnerable to insanity (Horne 1833, 37). According to Horne, critics almost always failed to recognise the true value of such men of genius, and publishers refused to publish their works, because 'it requires genius to discover genius' (Horne 1833, 105). Such a feat was far beyond the abilities of the anonymous readers employed by publishers, of whom Horne asserted that '[the reader's] grand climacteric is a "Commission of Lunacy" against Genius, he finds it so very unlike himself' (Horne 1833, 169). Nor was it only such readers who tend to dismiss genius as madness; and while Horne argued that the public would usually respond positively to works of genius if given a chance to see them, he also maintained that the general population would always tend to regard the geniuses who created such works as being insane.

Perhaps recalling Macaulay's remarks about the connections between poetry and 'unsoundness of mind' in his 1825 essay on Milton, which he later criticised at length in *A New Spirit of the Age* (1844), Horne declared that 'A man must always appear mad, the causes and effects of whose excitements are not understood'. He asserted that the mind of one genius could only be understood by another:

> It is as impossible for a man whose heart does not sympathise with that of his author, and whose imagination cannot meet him half-way, to enter into the feelings and thoughts of any great genius, as for the pin's-point eye of a mole to behold, with due admiration, the majesty of the mid-day sun[.] (Horne 1833, 122, 105)

For Horne, the perceptive abilities of the genius and non-genius have as little in common as those of a man and a mole—an idea he later explored in his novel-length fable, *The Poor Artist* (1850), in which different types of people turn out to see the world just as differently as different animal species. Following his premises through to their logical conclusions, Horne thus called for the establishment of a 'Society for English Literature and Art, &c', which would be staffed by men of acknowledged genius and funded at public expense. Given that only one genius could recognise another, the task of these geniuses would be to accurately identify other 'men of genius', who would then be enrolled in the society and financially supported by the public. Being thus freed from want, they would be able to devote themselves to the creation of great works of genius, untroubled by the demands of the marketplace or the incomprehension of the outside world (Horne 1833, 287–308).

The sincerity of Horne's *Exposition* cannot be doubted: and yet it took the Shelleyan concept of the misunderstood genius to such extremes that it risked sounding more than a little insane, especially during the extraordinary 50-page diatribe that Horne directed at the readers employed by publishers to judge the quality of new manuscripts. For the genius, it seems, everything means the opposite of what it appears to mean. If a genius does badly at school and at college, proves incompetent at business, and has their works rejected by publishers and savaged by critics and reviewers, then these are just so many proofs of the true superiority of their intellect and 'nervous system': and if most readers are unable to see any value in their works, then that just demonstrates how extraordinary their genius really is (Horne 1833, 37, 105, 119, 241). Horne articulates a paranoid vision of the world in which true geniuses almost never succeed during their own lifetimes, because the envious and uncomprehending non-geniuses are always engaged in a conspiracy against them: at times, such as when he speculates that the readers employed by publishers may have 'a secret influence with certain newspapers and inferior periodicals', he tips over into outright conspiracy theory (Horne 1833, 203n). The society of geniuses which he proposes would, to the non-genius, presumably look like a madhouse: a group of extremely peculiar people doing incomprehensible things for incomprehensible reasons, and gravely informing the public that if only they possessed sufficient genius they would recognise that all this apparently worthless activity was, in fact, of inestimable value, and entirely worthy of the generous public subsidy it received.

Given all this, it is hardly surprising that John Wilson, in a brutal review of Horne's book for *Blackwood's Edinburgh Magazine*, dismissed him as a lunatic: 'a case of Monomania well worthy the attention of Phrenologists', 'blind and deaf to all that is seen and heard by the sane'. Clearly enjoying himself, Wilson went on to recommend that Horne should be treated with purges, cupping, and shower-baths, to prevent his madness escalating into murder or suicide (Wilson 1833, 440–1). And yet the *Exposition* only articulated, in an exaggerated form, the same idea of artistic genius that Tennyson, Browning, and Hallam had derived from Shelley and Byron: that it was a 'birthright' which forever set its bearer apart from ordinary men and women, dooming them to be scorned and neglected during their lifetimes, but allowing them to sustain themselves amidst their suffering by their mystical communion with what Horne called 'The Fathomless' (Horne 1833, 133, 252). Horne believed that taking up such a vocation involved a real chance of being regarded as mad by one's contemporaries, and a not inconsiderable risk of actually succumbing to insanity under the weight of one's disappointments. But the true genius would always know their own worth, even if the world did not; and history, Horne argued, always vindicated their judgement in the end.

A more sober account of the ways in which poetic geniuses might be psychologically different to other people was offered by another member of Fox's circle, John Stuart Mill, in his 1833 essay 'Thoughts on Poetry and Its Variations'. The idea that poets are born rather than made had originated, Mill wrote, 'in the infancy of psychology', but could still be demonstrated to be basically true 'now when that science is in its adolescence':

> [T]here are poetic *natures*. There is a mental and physical constitution or temperament peculiarly fitted for poetry. This temperament will not of itself make a poet, no more than the soil will the fruit; and as good fruit may be raised by culture from indifferent soils, so may good poetry from naturally unpoetical minds. But the poetry of one who is a poet by nature will be clearly and broadly distinguishable from the poetry of mere culture. It may not be truer; it may not be more useful; but it will be different: fewer will appreciate it, even though many should affect to do so; but in those few it will find a keener sympathy, and will yield them a deeper enjoyment. (Mill 1981, 1:355)

For Mill, poetry had nothing to do with linguistic skill. Instead, he defined it as the authentic expression of human emotion, going so far as to assert that 'whosoever writes out truly any human feeling, writes poetry': but while this might suggest that anyone could write poetry, Mill argued that to be a true poet required a specific kind of 'mental and physical constitution'. Because such poets are people to whom 'nature has given strong feelings' and 'intense sensibility', the emotional associations which linked their ideas together would always be stronger than the rational ones, and as a result they would perceive a version of reality which would look very strange to other people: 'a phantom-world', 'a place of *ignes fatui* and spectral illusions' (Mill 1981, 1:356).

The poems produced by an individual with such a 'poetic temperament' would be works in which 'thought [waits] upon feeling', whereas in the works of the 'poet of culture', for whom the association of ideas along emotional rather than rational lines is a matter of conscious choice and deliberate mental effort, 'feeling waits upon thought'. For Mill, these two types were exemplified by Shelley and Wordsworth, respectively. Wordsworth was a great poet because poetry was the direction in which he consciously chose to direct his mental energy, but Shelley's poetic gifts were due to his unusual physical constitution, his 'impassioned nature', his 'original fineness of organisation', and the 'susceptibility of his nervous system'. Wordsworth could have 'chosen a different walk of authorship': but Shelley, with his super-susceptible nervous system, could only ever have been a poet (Mill 1981, 1:356–60).

There are obviously strong parallels between Mill and Hallam's accounts of the poetic character. For both of them, as for Horne, poetic ability is primarily a matter of mental organisation rather than verbal dexterity, and the explanation for why some people are capable of writing great poetry was to be sought for not in the biographies of poets but in their nervous systems. Both saw Wordsworth as an example of a poet whose poetry was subordinated to what Mill called 'thought' and Hallam called 'reflection', made to serve as a vehicle for moral or intellectual ideas rather than being allowed to function as an end in itself, and both saw Shelley as representing the opposite poetic tendency. Where Mill differed from Hallam was in seeing *emotion* rather than *sensation* as the quality in which the natural, Shelleyan poet would excel, with the result that his conception of poetry is even more intensely 'psychological' than Hallam's. For Hallam, the 'fine organs' of the poet include the eyes and ears; his 'poet of sensation', by virtue of this superior biological equipment, actually sees the world differently to other people. For Mill, however, the poet's 'fineness' is purely internal: a matter of perceiving the same sense-data as everyone else, but assembling it in a different order due to their unusual 'mental and physical constitution'.

Rather than seeing true poets as great communicators, blessed with the linguistic abilities needed to express their thoughts with exceptional clarity and force, Mill, Horne, and Hallam ultimately proposed models of poetry which had nothing to do with language at all. For them, the highest form of poetry resulted from a set of rare physical abnormalities within the brain and nerves of the poet, which led them to experience the world differently from other men and women. (In his 1835 review of Tennyson's poems, Mill diagnosed Tennyson as possessing exactly this 'peculiar kind of nervous susceptibility') (Mill 1981, 1:413). Precisely because the world they described was so different from that which most people perceived, the majority of people would never appreciate their poems, dismissing them as dreams, visions, or simple falsehoods. Writing to Carlyle in 1833, and possibly recalling Macaulay's 1825 assertion that 'Truth, indeed, is essential to poetry; but it is the truth of madness', Mill remarked that poetic truths would strike most people 'as nothing but dreaming or madness' (Mill 1981, 12:163; Macaulay 1843, 8).

All this demonstrates the extent to which Horne, Mill, and Hallam—and, by extension, the Fox circle and the Cambridge Apostles, Tennyson and Browning included—had, by the early 1830s, internalised a conception of the true poet as a visionary genius, able to perceive truths about the world which remained hidden from all others. But they articulated this conception in distinctly medicalised terms. Wordsworth had seen the poet's gifts as being due to their possession of 'a more comprehensive soul [than is] common among mankind': but Mill, Horne, and Hallam, like Isaac D'Israeli before them, located the source of the visionary capacity of poets not in their souls, but in their nerves (Wordsworth 1992, 751).

'Such Genius Is Itself a Disease': Madden and Croker on the Infirmities of Genius

Although Hallam, Horne, Mill, and D'Israeli all offered basically biological explanations for poetic genius, none of them described the physiological abnormalities which gave rise to poetic talent as being inherently pathological. The same, however, could not be said of the Irish physician Richard Madden, who in 1833 published precisely the kind of 'physiological history' of genius that D'Israeli had hoped for back in 1822. This work, *The Infirmities of Genius*, opened with these chilling words:

> It is generally admitted that literary men are an irritable race, subject to many infirmities, both of mind and body; that worldly prosperity and domestic happiness are not very often the result of their pursuits. (Madden 1833, 1:13)

Madden's thesis was, essentially, that all great writers tended to be more or less mad. Prolonged literary labour wore out the brains of writers, predisposing them to melancholy and hypochondria: a belief which, as Ruston has shown, had become increasingly widespread during the Romantic period (Ruston 2016, 299–301). 'Studious habits inordinately pursued' and 'mental exertion long continued' gave rise to 'the morbid sensibility of genius' which made writers into such peculiar people. Worse still, too much mental labour could cause a derangement of the nerves which connected the brain to the rest of the body, controlling physical functions such as digestion, and 'when once their action is deranged, the whole animal economy suffers from that derangement' (Madden 1833, 1:17, 1:40). These physical and mental problems were often aggravated by the fact that the mental exhaustion caused by literary work commonly led writers to 'the use of stimulants' such as alcohol and laudanum, which provided temporary relief but only served to worsen the underlying condition (Madden 1833, 1:23, 2:160). The ultimate result of all this nervous derangement and substance abuse was, frequently, an early death: Madden even compiled a comparative table of longevity to demonstrate that great poets died, on average, almost 18 years younger than great natural philosophers (Madden 1833, 1:68). Worst of all were those cases in which precociously gifted writers such as Tasso (or Tennyson) began their literary works early in life. As Madden noted, 'the sooner the mental powers are developed, the sooner the bodily powers begin to fail', and they usually died young. Their minds suffered, too: 'they commenced by being prodigies, and they ended by becoming stupid' (Madden 1833, 1:60, 1:63).

The consequence of all this, in Madden's view, was that the people who were best placed to understand the minds of great poets—or 'wayward child[ren] of genius', as he called them—were doctors, who were qualified to explain the nature and aetiology of their various mental pathologies (Madden 1833, 1:20; Ruston 2016, 304). (D'Israeli had hinted at a similar perspective in 1822, asserting that a non-genius such as himself could write about geniuses, in the same way that a doctor could write about patients whose illnesses he didn't share) (D'Israeli 1822, vii). The problem with people like Shelley and Byron, Madden implied, was that their biographies had not been written by medical men. Such medical biographers would have been able to sympathetically explain that their odd and antisocial behaviour had simply been the result of bodily illness, which had in turn been the almost inevitable consequence of the mental fatigues incurred during their pursuit of poetic excellence (Madden 1833, 1:16, 20–1). In fact, Madden maintained that increased medical knowledge always led to improved sympathy and compassion:

> The more extensive our knowledge of human nature is, and the better acquainted we make ourselves with that strong influence which mind and body mutually exert, the greater will be the indulgence towards the errors of our species, and the more will our affections be enlarged. (Madden 1833, 1:22)

In Madden's account, the tormented lives of poets such as Burns, Cowper, and Byron were straightforward cases of medical mismanagement. Their literary labours deranged and depleted their stock of nervous energy, and the lifestyles they adopted were ill-adapted to enable a recovery: they consumed too much laudanum and alcohol, ate the wrong food, lived in the wrong places, read the wrong books, and so on, with the result that their lives became downward spirals of physical and mental deterioration. If only a competent physician (such as himself) had been on hand to give them sound medical advice, then Cowper could have been saved from his almost life-long misery, and Burns and Byron from their untimely deaths. As it was, however, they deserved sympathy, not condemnation: they had not been bad people, merely sufferers from the physical and mental ailments which were inevitably attendant upon chronic brain fatigue.

Madden's book was very much in the tradition of such near-contemporary works as Samuel Hibbert's *Sketches of the Philosophy of Apparitions* (1825), Isaac Taylor's *Natural History of Enthusiasm* (1829), Walter Scott's *Letters on Demonology and Witchcraft* (1830), Thacher's *Essay on Demonology* (1831), and Brewster's *Letters on Natural Magic* (1832), insofar as it provided a reassuringly rational medical explanation for some very bizarre and disturbing behaviour. (Fittingly, Scott—who had died in 1832—was Madden's prime example of a literary genius who, through careful management of his health, had managed to keep himself reasonably sane until very late in life.) As Carlyle complained in *Sartor Resartus*, 'Witchcraft, and all manner of Spectre-work, and Demonology, we have now named Madness, and Diseases of the Nerves' (Carlyle 1975, 195). For Madden, the poet, like the mystic and the demoniac, was not to be judged in metaphysical or moral terms, but as a medical case study: and in his accounts the *Weltschmerz* of the Romantics shrinks to a purely biological problem, perfectly curable through proper management of one's nervous energy.

Like phrenology—which Madden invoked at one point to explain Cowper's monomania—and the rest of the new psychology of the period, *The Infirmities of Genius* was characterised by an impressive confidence in the ability of medical science to account for even the oddest forms of human behaviour, and a corresponding insistence that it was the doctor who, by virtue of his professional training, was best equipped to understand and deal with them (Madden 1833, 2:65–6). It thus formed part of the long-running cultural battle through which, as Corfield has described, the British doctors of the period sought to establish their professional authority in all matters of physical and mental well-being—often at the expense of other professions, as can be seen from Madden's lament at the fact that when Cowper was tormented by the belief that he was damned, he was sent to a clergyman for religious instruction, when he obviously should have been turned over to a doctor for medical treatment instead (Madden 1833, 2:23; Scull et al. 1996, 44–5; Corfield 1995, 165).

As part of their campaign for professional acceptance and recognition, many early nineteenth-century doctors made large claims for the good that could be accomplished if only people could be induced to place their trust in properly scientific forms of professional medicine (Scull et al. 1996, 5–6). If evil, crime, and religious enthusiasm were all essentially medical matters, then the medical profession, properly supported, could do much more than maintain the bodily health of the people of Britain: it could also safeguard them from falling into religious error, depravity, and vice. Madden was unusual only insofar as he asserted that the medical profession had yet another potential gift to offer to the nation, namely, the improved happiness, longevity, and productivity of British poets. He clearly intended his book to be an encouraging one, and in his final chapter he dropped some broad hints to the effect that he hoped any literary men who happened read it would follow his medical advice, and thus save themselves from the miserable fate of a Byron or a Cowper (Madden 1833, 2:188–9). Madden did not quite pathologise poetry itself, but he did make clear that the writing of poetry, like the drinking of gin or the eating of opium, was a very dangerous activity, and that an unwise over-indulgence in it could lead the incautious poet to the madhouse or the grave. Nor did his warnings go entirely unheeded: and when, in 1841, Benjamin Haydon heard of the death of his fellow artist David Wilkie, he recalled ruefully that he had 'begged [Wilkie] to read Madden, to understand the nature of the diseases' that, in Haydon's view, had ultimately caused Wilkie's death (Haydon 1853, 3:177).

Madden, of course, was hardly the first person to suggest a connection between poetic talent and madness. As Whitehead points out, the association goes back at least as far as Aristotle: the idea that 'the lunatic, the lover, and the poet / Are of imagination all compact' had been a cliché since Shakespeare's day, and I have already mentioned Macaulay's famous assertion that 'Perhaps no person can be a poet, or can even enjoy poetry, without a certain unsoundness of mind' (Whitehead 2017, 35–44; Macaulay 1843, 7; Shakespeare 1984, 116). However, while it was generally accepted that poets were often eccentric and prone to madness, not everyone agreed with Madden's thesis that great poetic achievements and great mental fragility necessarily went hand in hand. Milton and Shakespeare were generally regarded as having been exemplars of mental clarity: thus Horne and Barrett, in 1844, rebutted Macaulay's association of poetry with 'unsoundness of mind' on the grounds that it led 'directly to the question of whether Shakespeare and Milton ought to be regarded in any respect as lunatics', clearly assuming that the reputation of both writers as paragons of sanity would lead their readers to view such a question as absurd (Horne 1844, 2:40–1). Lamb asserted in 1826 that 'it is impossible to conceive of a mad Shakespeare', while Haslam's 1809 *Observations on Madness* argued that highly imaginative people often went mad, but denied any link between genius and delirium: true genius, he maintained, was not dependent upon the imagination (Felluga 2005, 28–9; Lamb 1954, 219). Conolly, in his *Inquiry Concerning the Indications of Insanity* (1830), cited both Milton and Shakespeare as examples of what a poet could achieve if his mind was so 'happily constructed' that 'his mighty imagination never overmastered his understanding', while dismissing 'the "great wits" who are "allied to madness"' as a distinctly inferior breed (Conolly 1830, 82–3, 128–9).

In a similar vein, in his 1833 review of Madden's book, Croker wrote in *The Quarterly Review*:

> [A]lthough we must admit that many men of genius have afforded cause to suspect that great wit is—as a great wit said—nearly allied to madness, and that we have often seen it connected with other bodily infirmities, and particularly scrofula, we do not venture to suggest that this is always or even generally the case; on the contrary, we repeat, that some of the greatest geniuses that ever adorned the world, also improved it by the purest examples of moral and mental rectitude—*mens sana in corpore sano!* (Croker 1834, 55)

Despite this, the fact that both Conolly and Croker alluded to Dryden's line—'Great Wits are sure to Madness near ally'd'—indicates the extent to which the idea that poets were often somewhat mad had gained common currency by the 1830s. In its original context, Dryden's line was actually an observation regarding the risks taken by political schemers, and yet both Croker and Conolly, like Lamb before them, apparently took it for granted that it could be applied with equal validity to poetic 'wits' as well (Dryden 1972, 2:10; Lamb 1954, 219).

While both Conolly and Croker agreed that madness (and possibly scrofula) often accompanied the practice of poetry, they insisted that genius did not require or result from madness. Both asserted that the very greatest poets, far from being the maddest and most scrofulous of men, tended to be models of physical and mental health (Conolly 1830, 129–31, 147–9, 187–8). (A similar claim was advanced by Heraud in 1830, who used it to argue that Byron couldn't have been a real genius, because if he had been then he wouldn't have been so sick and deranged) (Higgins

2005, 30–1). By the 1830s, however, the legend of Byron (and, to a lesser extent, Shelley) had popularised the idea that antisocial mental abnormality might, in a hero or artist, be a proof not of the inferiority of their genius but of its *superiority*. For Byron, the true poet was likely to be both the gloomiest and most brilliant of men; and Madden, who was clearly very sympathetic to Byron, simply attempted to explain this phenomenon in medical terms.

Over the course of his blisteringly hostile review of *The Infirmities of Genius*, Croker savaged Madden's book for a whole range of real and imagined failings: lack of philosophy, lack of Classical knowledge, incoherence, plagiarism, self-contradiction, neglect of women, inadequate statistical methodology, lack of propriety, insensitivity towards the recently dead, and much more besides (Croker 1834, 34–56). Yet even Croker conceded that men of genius, taken in the aggregate, really did tend to be a mentally fragile bunch:

> [P]ersons of a lively imagination, which is commonly called genius, may be more liable than ordinary persons to mental derangement, and of course to those species of bodily infirmities which are more peculiarity influenced by the mind. [...] We do not pretend to define what 'genius' is, but we think we may say, that, in its popular sense, it is generally, if not always, accompanied by, if it does not consist in, great mental sensibility, superior acuteness, and a more delicate susceptibility of impressions: such minds may be more liable to be deranged by sudden shocks, or impaired by over use, than less delicate organs. We recollect being very much struck with an observation which was made to us by the intimate connexion of one of the most illustrious men of our time, who,—after stating, in answer to our enquiry after our common friend, the wonderful activity of his mind, his acute sensibility, and the high pitch to which all his sensations were tuned,—added, 'He cannot be well; such genius is of itself a disease.' (Croker 1834, 51)

It is somewhat surprising to read this passage when it is remembered just how freely Croker himself dealt out 'sudden shocks' to such 'persons of a lively imagination' as Keats, Hunt, and Tennyson. However, it does serve to demonstrate that, by 1833, even someone as hostile as Croker to what would now be called 'Romanticism' was largely willing to accept a model of poetic talent that saw 'genius' as being due to the extreme and unusual delicacy of the poet's mental 'organs'—a view very similar to that advanced by Mill and Hallam in their respective defences of Shelley. As a result of this delicacy, such people were able to write great poetry, but their brains were also 'more liable to be deranged by sudden shocks, or impaired by over use'.

As I have mentioned, Croker did not believe that *all* poets suffered from this mental delicacy. He stressed that while such an individual '*may* be more liable than ordinary persons to mental derangement' and '*may* be more liable to be deranged by sudden shocks', equally they may not; and like Conolly, he believed that the very greatest of poets were usually made of sterner stuff (Croker 1834, 51, 55—my italics). But despite his many quarrels with Madden, Croker evidently agreed that a real connection existed between great imaginative talent and great mental fragility, and that in extreme cases the connection between the two could become so marked that one might justly say 'such genius is of itself a disease'. For Croker, in other words, Madden's assertion that some forms of poetic talent essentially had to be managed and treated as though they were forms of mental illness seems to have been one of his *least* objectionable claims. Whitehead has noted the importance of the periodical press in disseminating the medical psychology of the period to a general audience—as he writes, 'the early nineteenth century was perhaps the last period when discussion of the most advanced medical science generally took place across such a widely visible public sphere'—and given that Croker's review was almost certainly read by many more people than Madden's original book, it may well have played at least as strong a role as *The Infirmities of Genius* itself in propagating this idea among the British reading public (Whitehead 2017, 101).

'Almost Prophetic': J.C. Colquhoun and the Mesmeric Explanation of Poetic Vision

Madden's theory of psychology rested on a 'vital fluid' which pervaded the human body, and was transmitted via the nerves from one organ to another. He called this fluid 'nervous energy', but he remarked ruefully that neither he nor anyone else was really sure what it was:

> Although science (with all the rapidity of its march) has thrown little if any additional light on its phenomena for the last thirty years, yet a few facts have been noticed whose tendency is to show that there is a similitude between the phenomena of the nervous and electric fluids.

> Whenever the properties of the latter shall be better understood than they are at present, in all probability the principle of the nervous energy will be more cognizable to the range (limited as it must necessarily always be) of human knowledge. (Madden 1833, 1:47–8)

The belief that an invisible electric fluid was responsible for animating living tissue was hardly a new one: the electrical experiments carried out by Galvani in 1780 had seemed to imply as much, although, as Madden noted, this bioelectricity had proven frustratingly difficult to study (Fulford 2004, 60; Hughes 2015, 34–5). Madden himself hoped that Faraday, who had begun his researches into the properties of electricity in 1831, might be able to provide some new insights into how electricity made life and thought possible. As it turned out, Faraday did no such thing, but over the course of the 1830s he *did* contribute to the growing accumulation of evidence that pointed towards some intimate relationship between electricity and the equally mysterious force of magnetism (Matthew and Harrison 2004, 19:30–2). This evidence, limited though it was, was enough to help prepare the way for the 'science' of mesmerism, which set out to solve exactly the problem that Madden had stated: how the galvanic forces and magnetic and electrical 'fluids' pervading the human body related to the functioning of the human nerves and brain.

Originally developed by Franz Mesmer in the 1770s, mesmerism had received a rocky reception in eighteenth-century Britain, where it had been perceived as tainted by its association with questionable science and even more questionable politics (Fulford 2004, 64–6; Hughes 2015, chap. 1). In France and Germany, however, it had flourished, swiftly developing from a simple method for healing disorders of the mind and body into a technique for inducing higher states of consciousness. By the 1830s, patients placed in mesmeric trance states had demonstrated the apparent abilities to foretell the future, see through solid objects, perceive events taking place many miles distant, and many other feats besides (Fulford 2004, 66; Winter 1998, 122–4, 143–6). When Baron Dupotet arrived in Britain in 1836 and began to offer mesmeric training to his students, he thus came as the populariser of a much more ambitious and self-confident version of mesmerism than the one which had withered there in the 1790s, reinforced by a half-century of experimentation and accumulated evidence, and eager to assert its ability to explain the reality of seemingly supernatural events in scientific terms. Dupotet's lectures helped to lay the groundwork for a British mesmeric revival, the leading advocates of which included the lawyer J.C. Colquhoun, the writers Harriet Martineau and Chauncey Hare Townshend, the doctor John Elliotson, and—most famously of all—the novelist and journalist Charles Dickens (Hughes 2015, 10, 82–3).

Madden does not seem to have been a believer in mesmerism, and his summary of previous investigations into the nature of 'nervous energy' pointedly failed to mention Mesmer and his disciples (Madden 1833, 1:47–56). Such scepticism was common amongst British doctors of the period, and advocates of mesmerism repeatedly complained of the unwillingness of the British medical profession to give what they considered to be a fair hearing to the theory and practice of 'animal magnetism'. For mesmeric enthusiasts such as Colquhoun, the obvious resemblance between mesmerism and older magical practices, including the astral magic which Mesmer himself had studied in his youth, did not reflect negatively upon the credibility of animal magnetism. In their view, Mesmer had provided nothing less than a scientific explanation for the seemingly supernatural feats performed by the healers and mystics of previous ages: tales of miraculous cures, laying on of hands, visions of distant events, and accurate predictions of the future could all now be understood as examples of the power of animal magnetism (Hughes 2015, 54; Winter 1998, 249, 269; Barrow 1986, 77). The druids had practised a form of mesmerism; so had the 'ancient thaumaturgists', the witches of medieval Europe, and the healers of pagan Greece. Joan of Arc, Colquhoun asserted, had clearly possessed a knack for self-mesmerisation, which explained the visions that she experienced; the same was true of Jacob Boehme and Emmanuel Swedenborg (Colquhoun 1836, 2:102; Townshend 1854, 172–5; Crowe 1848, 1:37–9, 1:72–3; Sandby 1844, 232–62). Colquhoun even claimed that the Catholic church was deliberately trying to suppress the knowledge of mesmerism, as once it became generally known everyone would realise that the exorcisms and healing miracles from which the church derived so much income were simply mesmeric treatments (Colquhoun 1836, 2:187n).

Colquhoun's emphasis on the explanatory power of mesmerism was such that his book on the subject, *Isis Revelata* (1836), ultimately became a kind of mirror-image inversion of such contemporary works as Scott's *Letters on Demonology* or Taylor's *Natural History of Enthusiasm.* All three books gave what their authors considered rational, scientific explanations for seemingly supernatural phenomena: but while Scott and Taylor argued that modern medical medicine demonstrated why people might believe in such phenomena even though they did not actually happen, Colquhoun claimed that many of them *did* happen, and that modern medicine, in the form of mesmerism, had finally provided a scientific explanation for how such extraordinary events could take place. Nowhere is this reversal clearer than in the opposing explanations offered by Taylor and Colquhoun for the ongoing survival of the Catholic church. Both asserted that the Catholic clergy were secretly using their mastery of the psychological sciences to maintain their power over their gullible followers: but while Taylor saw them as using their mastery of psychology to persuade the faithful that miracles really happened (even though they didn't), Colquhoun claimed that they were in fact using their mastery of mesmerism to make those 'miracles' happen in the first place, and then deceitfully claiming that the resulting phenomena were carried out by divine intervention, rather than being the results of a medical technique which could be learned by almost anyone (Taylor 1830, 50; Colquhoun 1836, 2:187n).

In much the same vein, just as D'Israeli and Madden had asserted that modern medical knowledge could help to explain the nature of poetic genius, so Colquhoun implied that poetic inspiration could also be understood in mesmeric terms. In *Isis Revelata*, he wrote:

> [Mesmerically-induced somnambulism] is a state, as my readers must now be aware, totally different from that of ordinary life—a state in which the animal sensibility undergoes an essential change—a state in which the ordinary activity of the corporeal faculties is suspended for a time, and the internal instinct—the immaterial principle—perhaps the soul itself—displays its unfettered energies, independent of the material organs;—a state of existence which has been almost prophetically, although unintentionally, described, in the following beautiful line of one of the most philosophical of our living poets;—a state,
>
> In which the burthen of the mystery,
> In which the heavy and the weary weight
> Of all this unintelligible world,
> Is lightened:—that serene and blessed mood,
> In which the affections gently lead us on,—
> Until, *the breath of this corporeal frame*
> *And even the motion of our human blood*
> *Almost suspended*, WE ARE LAID ASLEEP
> IN BODY, AND BECOME A LIVING SOUL:
> *While with an eye made quiet by the power*

> *Of harmony, and the deep power of joy,*
> WE SEE INTO THE LIFE OF THINGS.

> It cannot fail, I think, to be perceived, with what remarkable, what wonderfully minute accuracy, this poetic description tallies with the phenomena which actually occur in the magnetic somnambulism. (Colquhoun 1836, 2:107–8—capitals and italics in original)

Colquhoun proceeded to compare each step of Wordsworth's account of his 'serene and blessed mood' to the experience of 'magnetic somnambulism', noting that each involves a state of forgetting, followed by the suspension of 'the ordinary functions of the animal organization', a strengthening of 'the internal power of vision', and finally an experience of clairvoyance (1836, 2:108–9). 'Almost prophetically, although unintentionally, described …' Poor Wordsworth! He thought he was actually describing his own visionary insights—but, in fact, he had just been 'unintentionally' describing 'the phenomena which actually occur in the magnetic somnambulism' all along!

For Colquhoun as for Madden, then, the mental peculiarities of poets were ultimately explicable in medical terms. For Madden, poets who experienced visions and strange states of consciousness were suffering from a derangement of their nervous energy: for Colquhoun, they had unknowingly mesmerised themselves into somnambulistic states through the unintentional manipulation of their own magnetic fluids, just like Swedenborg, Boehme, and Joan of Arc. Colquhoun differed from Madden insofar as he maintained that such altered states could produce real and seemingly supernatural insights, up to and including prophetic knowledge of the future, but both concurred in seeing poets themselves as unqualified to understand their own altered states of consciousness. What they needed was an expert—a medical doctor for Madden, a skilled mesmerist for Colquhoun—who could explain such states to them, and help them to minimise the amount of mental harm that they inflicted on themselves in the course of their poetic endeavours. As Madden's grim case histories made all too clear, they would only make a fearful mess of things if left to their own devices.

Vision and Delusion in Tennyson's Early Poetry

The relationship between vision and delusion, poetic inspiration and insanity, seems to have preoccupied Tennyson from a very early stage in his career. As a young man he was strongly drawn to the ideal of the poet as a supernaturally powerful seer or magus, and his 1830 poem 'The Mystic' was devoted to the rapturous description of a Romantic visionary *par excellence*: one whom 'Angels have talked with', but whom the ordinary people have treated 'with an undiscerning scorn' because 'he was not one of ye' (Tennyson 1987, 1:252). Tennyson's 1830 collection *Poems, Chiefly Lyrical* contained two more poems, 'The Poet' and 'The Poet's Mind', which presented similarly idealised depictions of the poet as visionary genius, one who 'saw through life and death, through good and ill' and whose thoughts were 'winged with flame' (Tennyson 1987, 1:243–4). Like 'The Mystic', these works insisted that the true poet would inevitably be set apart from others by the magnitude of their gifts, which lesser mortals could never truly comprehend:

> Vex not thou the poet's mind
> With thy shallow wit:
> Vex not thou the poet's mind;
> For thou canst not fathom it. (Tennyson 1987, 1:246)

The young Tennyson knew a thing or two about unfathomable mental states himself, and in 1838 he wrote to Emily Sellwood that 'I have dim mystic sympathies with tree and hill reaching far back into childhood' (Tennyson 1982, 1:166). From an early age he had experienced involuntary trance states; he had also experimented with deliberately inducing them, repeating his own name over and over as a form of self-hypnosis. Years later he wrote that he had 'often had a strange feeling of being bound and wrapped in the Great Soul', a feeling which, if it could be trusted, would elevate his trances to the level of genuine religious or mystical experiences. But such events could just as easily have been classified as symptoms of mental pathology rather than signs of inspiration, and he accordingly kept his trance states secret until 1848 (Faas 1988, 54–5, 74; Nichols 1986, 132–3; Tennyson 1973, 188).

As Colley and Herb Wright have shown, Tennyson suffered from persistent worries about his own apparent vulnerability to lunacy, as well as the anxiety that his trances might be symptoms of epilepsy, inherited from his father—a disease which, at the time, was believed to often lead to madness (Colley 1983, 3–8, 51; Herb Wright 1987, 61–76). 'I have studied the minds of my own family,' he wrote to his uncle in 1834: 'I know how delicately they are organised' (Faas 1988, 54). 'Vex not the poet's mind; / For thou canst not fathom it', he had written in 1830: yet his younger brother Edward had also been a poet, one who had apparently 'fancie[d] himself to be a superior Genius', and Tennyson had seen his mind grow so very unfathomable that, in 1832, he was carried off to a private lunatic asylum in York (Tennyson 1982, 1:29). What if that was to be his fate, also? What if his trance states and mystical experiences were not moments of visionary insight, but simply symptoms of physical infirmity or incipient insanity?

One early work in which Tennyson appears to have wrestled with these problems is his 1829 poem, 'Timbuctoo'. This was itself a rewrite of the most ambitious of his adolescent poems, 'Armageddon': an unfinished epic fragment in Miltonic blank verse, which he wrote when he was about 15. 'Armageddon' had purported to describe a visionary experience, in which the first-person speaker is visited by a 'seraph' as he sits overlooking the valley of Megiddo. 'Open thine eyes and see!', the angel tells him, and blesses him with the power of divinely enhanced vision:

> I felt my soul grow godlike, and my spirit
> With supernatural excitation bound
> Within me, and my mental eye grew large
> With such a vast circumference of thought,
> That, in my vanity, I seemed to stand
> Upon the outward verge and bound alone
> Of God's omniscience. Each failing sense,
> As with a momentary flash of light,
> Grew thrillingly distinct and keen. I saw
> The smallest grain that dappled the dark Earth,
> The indistinctest atom in deep air,
> The Moon's white cities [...]. (Tennyson 1987, 1:81)

The angel grants to him, in other words, precisely the kind of heightened perception and prophetic visionary power which had been celebrated by Shelley and Wordsworth, and which Tennyson would go on to celebrate, in turn, in 'The Mystic', 'The Poet', and 'The Poet's Mind'. During his time at Cambridge, Tennyson rewrote this poem for submission to the 1829 Chancellor's Prize, for which the set topic that year was 'Timbuctoo'. In the rewritten version, the speaker is still visited by an angel on a mountaintop, although this time he is standing on the Rock of Gibraltar wondering about Timbuktu rather than standing near Mediggo wondering about Armageddon. He is once again granted a form of visionary perception which is described in the same terms, albeit with slightly reduced levels of hyperbole: he feels his soul grow 'mighty' rather than 'godlike', and seems to stand on the verge of 'full beatitude' rather than 'God's omniscience'. In this happy state he is able to transcend the distances which separate him from Timbuktu, and see its 'wilderness of spires, and chrystal pile / Of rampart upon rampart, dome on dome'. But then something rather extraordinary happens, which has no parallel in the earlier 'Armageddon' poem. The vision-granting 'Seraph' identifies itself as 'the Spirit ... of Fable', and warns him that his fantastical vision of the legendary city is not actually true. The real Timbuktu is just a 'Low-built, mud-wall'd, Barbarian settlement', and the 'wilderness of spires' which he has seen has no existence outside his own mind (Tennyson 1987, 1:193–8).

This change has some far-reaching consequences. What in 'Armageddon' had been presented as genuine divine revelation, granting the speaker unmediated access to a higher level of reality, becomes in 'Timbuctoo' nothing more than a 'Fable'. When the speaker in 'Armageddon' turns his 'mental eye' upwards and sees 'the Moon's white cities', it is because those lunar cities are really there, waiting for an eye clear enough to see them; but when the speaker in 'Timbuctoo' does the same, the cities he sees, like 'chrystal' Timbuktu itself, are purely imaginary. To be visited by an actual seraph is to undergo a life-changing religious experience—in the Bible, it is an encounter with the seraphim which begins Isaiah's career as a prophet—but to be visited by a seraph who is actually the Spirit of Fable is essentially just an elaborate way of describing an imaginative reverie. As Tucker puts it, 'The Spirit of Fable in "Timbuctoo" bears a message about fictiveness that in effect takes his seraphic self out of commission and sets the poet up on his own' (Tucker 1988, 70). This does not render the resulting experience valueless, as the Spirit himself explains: indeed, he asserts that 'There is no mightier Spirit than I to sway / The heart of man', because the fables of which he is the embodiment have the power to 'teach him [i.e. 'man'] to attain / By shadowing forth the Unattainable' (Tennyson 1987, 1:197). Fictional fables can teach real lessons: but, despite this, the spires and cities revealed by such fables remain shadows and not realities.

This, in turn, does not need to be a problem, provided we can be sure of recognising such fables for what they are whenever we happen to encounter them. But on this, 'Timbuctoo' is far from reassuring: its speaker is every bit as caught up in his seraphic vision as the speaker of 'Armageddon' (as well he might be, given that the relevant sections of the two poems are almost identical), and it is only when the seraph reveals his true identity as the Spirit of Fable that the speaker realises that he has not actually been carried to 'the outward verge ... of full beatitude' after all. Faas rightly comments that in the transition from 'Armageddon' to 'Timbuctoo', 'Romantic vision quickly degenerates into hallucination', and 'a romantic dream of ideational absolutes ... has turned into a spectral vision as analysed by contemporary mental science': but what needs to be recognised here is just how close this 'spectral vision' comes to passing itself off as a genuine sacred revelation (Faas 1988, 74–6). The poem even provides a glimpse of what can happen when fable and reality, shadow and substance are confused, in a scene in which the speaker imagines a priestess kneeling before an idol in ancient Greece:

> Before the awful Genius of the place
> Kneels the pale Priestess in deep faith, the while
> Above her head the weak lamp dips and winks
> Unto the fearful summoning without:
> Nathless she ever clasps the marble knees,
> Bathes the cold hand with tears, and gazeth on
> Those eyes which wear no light but that wherewith
> Her phantasy informs them. (Tennyson 1987, 1:191–2)

The 'deep faith' of the priestess allows her to see light in the stone eyes of her god, which she interprets as a genuine sacred vision, revealing to her the hidden divinity which inhabits it: but, in fact, this is merely her 'phantasy' at work. The myths which have been woven around the 'awful Genius' she worships are, presumably, fables which were supposed to 'teach [her] to attain / By shadowing forth the Unattainable', and had she been able to recognise them as fables, she could perhaps have used them as an effective guide to her own life, and might now have a better response to 'the fearful summoning without' than hugging a statue's knees. But, instead, the reverse has happened: she has interpreted the fables as literal truths, and as a result she has wasted her life in the service of a lump of stone.

This, I think, is the anxiety which hovers behind 'Timbuctoo': the fear that the visionary truths we hold most dear may, in fact, be fables or delusions rather than realities, and that we may thus be misled into wasting our lives worshipping false gods, or searching for a 'chrystal pile' in lands where, in reality, there is nothing but a 'Barbarian settlement' to be found. What I want to emphasise, however, is the role that the priestess' imagination has played in bringing her to this point. Her faith is 'deep', but it is not blind; she does not worship the statue simply because she has been told to, but because she is convinced that she can actually see the divine light in its eyes, which for her constitutes proof that it is, indeed, inhabited by the god it represents. She sees the light because of the power of her 'phantasy'; had she been less imaginative (less poetic, one might say), she might not have been so easily misled. Imagination can be deceitful, and as a result, if, like the Shelleyan poet described in 'The Mystic', you are the sort of person whom 'Angels have talked with', then it becomes extremely important to be able to distinguish reliably between Spirits of Fable and genuine seraphs. Otherwise, disappointment or madness may await.

'Timbuctoo' won the prize for which it was submitted, and an anonymous review in the *Athenæum* declared that it marked Tennyson out as the true successor to the great poets of the previous generation:

> We have accustomed ourselves to think, perhaps without any very good reason, that poetry was likely to perish among us for a considerable period after the great generation of poets which is now passing away. The age seems determined to contradict us, and that in the most decided manner; for it has put forth poetry by a young man, and that where we should least expect it—namely, in a prize poem. (*The Athenaeum and Literary Chronicle* 22 July 1829, 456)

The review singled out for praise the section quoted from above, which describes the angel's descent and the vision which follows it, presumably feeling that these lines best illustrated the then-unknown author's ability to follow in the footsteps of his visionary predecessors in 'the great generation of poets which is now passing away'. But the poem remains an unsettling one, and not all its readers shared the enthusiasm of the *Athenæum*'s reviewer. Wordsworth's nephew Charles, who was himself a prize-winning student poet at Oxford at the time, declared that had it been submitted at Oxford rather than Cambridge it would not have been awarded a prize. It was more likely, he opined, that its author would have been encouraged to admit himself to a lunatic asylum instead (Tennyson 1987, 1:189).

Why should a poem such as 'Timbuctoo' mark its author out as being at risk of madness? It is not a 'mad' poem in the same way as, say, Smart's 'Jubilate Agno', insofar as it does not make statements which seem obviously bizarre or factually untrue. However, it does describe states of trance and vision, in which the speaker claims to have had distant and hidden things revealed to them, only to then assert that they are merely the 'shadows' conjured by the power of the speaker's imagination, and Tennyson himself described the poem as a 'wild and unmethodized [sic] performance' (Tennyson 1982, 1:109). This was the sort of material which made contemporary writers such as Macaulay, Madden, and Conolly fear that poetry might be psychologically unhealthy: it encouraged poets to dwell upon fables, indulging in imaginative fantasies in which the crucial distinction between fiction and reality could all too easily be lost, as they are in 'Timbuctoo' for the priestess and, for most of the poem, the speaker as well. It was for this reason that Conolly, in his 1830 *Inquiry*, warned his readers that 'Imagination' was 'a faculty all the varieties and all the influences of which the pathologist, but more especially the mental pathologist, ought to make his study' (Conolly 1830, 77).

Pauline: The Poet as Psychologist

A similar exploration of the difficulties involved in distinguishing between pathological delusion and poetic inspiration can be seen in Browning's first published poem, the dramatic monologue *Pauline* (1832). The speaker of *Pauline* resembles Browning in at least two important respects: he is a devoted admirer of Shelley and the tradition of visionary poetry that he represents, and he is interested in using poetry as a vehicle for psychological analysis. He declares:

> I strip my mind bare—whose first elements
> I shall unveil—not as they struggled forth
> In infancy, nor as they now exist,
> That I am grown above them, and can rule them,
> But in that middle stage when they were full,
> Yet ere I had disposed them to my will;
> And then I shall show how these elements
> Produced my present state, and what it is. (Browning 1983–2009, 1:42)

Like Tennyson in 1828–1830, the speaker of *Pauline* has an extremely high opinion of poets in general, and of Shelley in particular: he declares that the poet's words have the power to 'move / Men like a swift wind', and it seems to have been simply by writing poems that he once hoped to create a world in which 'Men were to be as gods'. He claims that it is only his joy at the fact that people were finally starting to appreciate the power and value Shelley's poetry which has saved him from utter despair: 'the glow I felt at HIS award / Assured me all was not extinct within'. For him, Shelley is the 'Sun-treader', a 'spirit', and, at one point, simply 'HIM': he is convinced that 'one so pure as thou / could never die', and asserts that he would give up all 'The wreaths of fame which seemed o'erhanging me, / To have seen thee, for a moment, as thou art' (Browning 1983–2009, 1:64, 1:56, 1:34–8).

The speaker is clearly a rather Shelleyan character himself, or at least fancies himself to be. He writes repeatedly of his yearnings for transcendence and infinity, his certainty that 'this earth is not my sphere', and his 'strange powers and feelings, and desires ... Referring to some state or life unknown'. He asserts that the one mental quality 'Which marks me'—presumably in the sense of marking him out from other people—is his imagination, which 'has been an angel to me': not a doubtful angel who comes 'In fitful visions', like the one in Tennyson's 'St Simeon Stylites', but a true angel which remains 'beside me ever, / And never failing me' (Browning 1983–2009, 1:70–4, 1:44). Despite possessing all these mental qualifications, however, the speaker of *Pauline* does not consider himself to be even remotely equal to Shelley, whom he believes to have been so very superior a spirit that even a poetic soul such as himself must regard him as a being of quite another order:

> For I have nought in common with him—shapes
> Which followed him avoid me, and foul forms
> Seek me, which ne'er could fasten on his mind[.] (Browning 1983–2009, 1:38)

Shelley thus becomes the yardstick by which the speaker measures himself and finds himself wanting. He believes that he cannot be like Shelley because he does not have a *mind* like Shelley—and, as a result, he is afflicted with doubt and weakness, quite unlike the godlike 'spirit' which he believes Shelley to have been.

This is an important qualification. There is no contradiction in poets acting both as amateur psychologists and as inspired, perfect visionaries, but there *is* a tension between the idea of the poet as psychologist and the poet as *flawed* visionary, plagued with mental 'foul forms' that testify to their imperfect control over their own mind—as the kind of inspired but unstable genius, in other words, that was so frequently described in the medical literature of the 1830s. This is very much the kind of genius that the speaker of *Pauline* presents himself as being, and his claims to be able to objectively analyse the nature of his own mind are thus undermined by the obvious lack of objectivity with which he writes about his past and present mental experiences. Recalling his past ambitions to build a world in which 'Men were to be as gods, and earth as heaven', he declares: 'I shall go mad, if I recall that time' (Browning 1983–2009, 1:56). Sure enough, as he explains his failure to achieve this ideal, his monologue becomes increasingly incoherent, breaking up into fragments, littering the text with ellipses, dashes and asterisks:

> I cherish prejudice, lest I be left
> Utterly loveless—witness this belief
> In poets, tho' sad change has come there too
> No more I leave myself to follow them:
> Unconsciously I measure me by them.
> Let me forget it: and I cherish most
> My love of England—how her name—a word
> Of her in a strange tongue makes my heart beat!
>
> ****
>
> Pauline, I could do any thing—not now—
> All's fever—but when calm shall come again—
> I am prepared—I have made life my own—
> I would not be content with all the change
> One frame should feel—but I have gone in thought
> Thro' all conjuncture—I have lived all life
> When it is most alive—where strangest fate
> New shapes it past surmise—the tales of men
> Bit by some curse—or in the grasps of doom
> Half-visible and still increasing round,
> Or crowning their wide being's general aim … (Browning 1983–2009, 1:78)

This is the earliest surviving example of a poetic technique which would become one of Browning's trademarks: the communication of a speaker's mental state through the form and syntax of their language. As Tate notes, 'Browning uses the poem's language to reveal more about the speaker's mental state than the speaker himself intends' (Tate 2012, 36). 'All's fever' here, indeed; the fragmented text mirrors the fragmented mind of the speaker, his lurching from topic to topic marked by the dashes and asterisks which mark the discontinuities of his speech upon the page. It is not clear what the speaker's 'belief / In poets' has to do with his 'love of England', except insofar as both are apparently examples of his 'prejudice', and towards the end of this passage he becomes almost totally incoherent. (I *think* his claim is that, by mentally rehearsing 'tales of men / Bit by some curse' and other stories of human life under extreme conditions, he has 'gone in thought' through 'all life / When it is most alive', and has thus already mentally experienced much more than 'the change / One frame should feel' in an ordinary human lifetime, but the sense is extremely obscure.) All this fragmentation serves to demonstrate that the speaker is in a highly agitated mental state, afflicted by 'wild fancies', but with the help of the addressee's sympathetic listening he ultimately pushes through them, returning to hope and coherence with the declaration that 'I am inspired—come with me, Pauline!' (Browning 1983–2009, 1:78–80).

Is the speaker inspired? He initially presents himself as someone who is able to pronounce authoritatively on his own mental states, due to both the extent of his psychological knowledge and his status as a poetic genius. As he reveals more of his mental infirmities, however, he runs the risk of shifting from being a psychological authority to a psychological case study, whose 'fever' and near-madness may need to be analysed and evaluated by someone else, as he seems too swept up in them to objectively evaluate them himself. When he speaks as a psychologist, he *states* what he believes to be the truth about his own mental states; but when he speaks as an 'inspired' poet, he *reveals* the truth about his mental states through the tone and syntax of his language, leaving the task of interpreting them to the reader. This doubleness, whereby the speaker is simultaneously presented as an authoritative psychological investigator *and* as a mentally imbalanced individual requiring authoritative psychological investigation, very much reflects the contradictory status assigned to such 'inspired' poets in contemporary British culture. As I have discussed, the abnormal and fever-prone brains of poets were thought to grant them both unusual levels of psychological insight and uncommon levels of mental instability, making them simultaneously the most and least authoritative of writers. The result is a tension between the two roles that the speaker aspires to: that of objective psychologist and that of wild, 'inspired' poetic genius.

As I have discussed, other members of the Fox circle, including Horne and Mill, held similar views on poetry to the ones expressed in *Pauline*: so similar, in fact, that the only favourable review that the poem received was the one published in Fox's *Monthly Repository* (Kelley et al. 1984–2016, 3:341–4). As Tate notes, '[*Pauline*] manifests an intellectual affinity with Fox's psychologically informed criticism, a shared interest in the possibility of utilizing poetry as a means of scrutinizing the mind' (Tate 2012, 31).

Horne and Mill associated poetic talent with both insight and madness, and identified Shelley as the most poetical of all poets, just as the poem's speaker did: Horne even shared the speaker's exalted views on the power of poets to change the world. Indeed, *Pauline* probably reflects what Browning, under the Fox circle's influence, imagined that a truly poetic character would be like: highly imaginative, somewhat incoherent, and very much unlike other people. Perhaps the most remarkable fact about the way in which *Pauline* depicts the poetic character of its speaker, however, is that he never mentions actually writing any poetry. For him, as for Horne and Mill, being a poet seems to be entirely a matter of having a certain highly unusual sort of mind: and his belief that poetic fame can be his, along with all the power that comes with it, seems to be based wholly on the fact that he has 'strange powers and feelings' and a strong imagination.

For the speaker of *Pauline*, being a great poet thus has nothing to do with developing his skills as a writer: instead, it is purely a matter of attaining a certain kind of 'inspired' mental state. Just as the speaker seems to believe that simply *thinking* about life under extreme conditions entitles him to claim that 'I have lived all life / When it is most alive', so he seems convinced that simply thinking grand thoughts is proof that he can become a great poet whenever he feels like doing so (Browning 1983–2009, 1:78). One consequence of this is that his poetic autobiography is played out almost entirely within his own mind, and involves only the most minimal interactions with the outside world: even Pauline herself seems to exist only to give him someone to monologue at. When *Pauline* was published anonymously in 1833, it was this intense inwardness—amounting almost to solipsism—which led Mill to famously conclude that *Pauline's* speaker was 'possessed with a more intense and morbid self-consciousness than ever I knew in any sane human being' (Browning 1983–2009, 1:11–12; Martens 2011, 29).

Along with imagination, intense self-consciousness is, in fact, one of the things on which the speaker of *Pauline* most prides himself:

> I am made up of an intensest life,
> Of a most clear idea of consciousness
> Of self—distinct from all its qualities,
> From all affections, passions, feelings, powers;
> And thus far it exists, if tracked in all,
> But linked in me, to self-supremacy [...] (Browning 1983–2009, 1:42–4)

It is not obvious how the speaker can be sure that his life is any more intense, or his idea of consciousness any more clear, than anyone else's: and as he gives no sign of ever having taken any real interest in the inner lives of anyone else, least of all that of his supposedly beloved Pauline, his confident assertion of his own superiority seems to rest on extremely slender foundations. (By 1840 Browning was more self-aware, noting wryly in *Sordello* that 'all men think all men stupider than they') (Browning 1983–2009, 2:336). The speaker clearly regards his possession of 'intensest life' as another proof that he is qualified to be an inspired visionary poet, but Mill read it less sympathetically, as a sign of self-involvement so intense that it risked drifting beyond the edges of sanity. At least one contemporary reviewer, who described *Pauline* as the work of a 'Mad Poet', evidently thought that its speaker might have crossed those borders already (Martens 2011, 32; Whitehead 2017, 125).

As Martens emphasises, the self-obsessed speaker of *Pauline* should by no means be straightforwardly identified with Browning himself (Martens 2011, 26). In fact, *Pauline* is a remnant of a plan which Browning dreamed up in 1832, but never put into practice: to anonymously publish works in several different forms, apparently by several different authors, and to allow each of these fictitious writers to attain some degree of literary renown, before finally revealing himself to be the author of all of them and basking in the resulting astonishment and fame (Kennedy and Hair 2007, 40). It is probably fortunate that this rather hare-brained scheme was never seriously undertaken, although it does point to a very early fascination on Browning's part with the creation of fictional alter egos—for the whole plan hinged on the works of each of these imaginary authors being sufficiently distinct that it would not be immediately obvious that they all issued from the same pen.

There are two such alter egos at work in *Pauline*, and Browning himself stands at least two removes distant from the text, having created the fictional Pauline who has supposedly edited this poem by her equally fictional lover. The note which Browning affixed to its text, supposedly written by 'Pauline' herself, is quite the most remarkable part of the poem. In it, she admits that much of *Pauline* doesn't make much sense, referring to it as 'cet étrange fragment' and as 'songe et confusion'; she also hints that maybe its author is just as confused by it as she is, declaring of its conclusion that 'Cette idée que je ne saisis pas parfaitement lui est peutêtre aussi intelligible qu'à moi'. The reason *Pauline* is so odd, she explains, is because of its author's mental peculiarity, his 'tournure d'esprit toute parliculière'; but, for reasons of her own (which she does not disclose), she declares that she has decided to preserve it anyway. As she helplessly concludes, 'La mieux serait de bruler ceci; mais que faire?' (Browning 1983–2009, 1:89).

As Armstrong argues, by attaching this note to the poem, Browning signalled to his readers that it should be read with scepticism rather than accepted at face value. Pauline's commentary, she writes, 'casts doubt upon the credibility of the text' (Armstrong 1996, 1107). By adding the voice of Pauline herself to the text, Browning allowed himself to lay claim to both sides of the argument: he was able to ventriloquise the half-mad visionary poet, all inspirations and incoherencies, while also maintaining an ironic distance from him, demonstrating that *he* understood the fallibility of such 'inspired' perspectives, even if the poem's speaker did not. Pauline's coy hint that she hasn't burned the poem for unspecified 'autres [raisons] encore plus puissantes' even raises the possibility that she only puts up with the speaker and his poetry for reasons of which he, who takes her reverence for his genius so thoroughly for granted, may be entirely unaware—reasons rooted in the external world of life and social relations to which he pays so very little attention.

But while the speaker of *Pauline* is not quite identical with the young Browning, he isn't all that different, either. As Browning himself acknowledged, of all his planned literary alter egos, this one—'the *Poet* of the batch'—was to 'have been more legitimately *myself* than most of the others' (Kennedy and Hair 2007, 40). Like the young Tennyson, Browning aspired to emulate the poetic achievements of Wordsworth, Shelley, and Keats: and everything *Pauline's* speaker has to say about his dreamy, bookish childhood, his fascination with poetry, his love of Shelley, and his interest in the workings of the human mind also seems to have been true of Browning himself. As a result, *Pauline* wavers between two states. On the one hand it is an almost-true, almost-sincere confessional poem about Browning's own state of mind at the age of 20, but on the other it is a psychological study in the form of a dramatic monologue, an exercise in imagining what it might be like to be a terribly sensitive, terribly imaginative, and rather mentally unstable young poet in 1833. Like the 'Victorian double poems' described by Armstrong, of which this is one of the earliest examples, it speaks with a double voice (Armstrong 1996, 12–13). It is at once the work of an earnest young writer who desperately wants his claims of inspiration to be accepted at face value, *and* the work of a more sceptical author who is well aware of just how fallible such would-be visionaries could be.

Despite these subtleties, Mill's criticism of the poem appears to have struck a nerve, and Browning went on to largely abandon the introspective mode of poetry which characterises *Pauline*. As Rivers notes, 'Browning realized from Mill's criticism that his analysis of subjectivity would have to be divorced from any suggestion of a confession' (Rivers 1976, 3). He later came to regard the poem as an embarrassment, keeping the unsold copies of it hidden in his attic, and refusing to allow Elizabeth Barrett to read it: it was, he explained, 'ambiguous', 'feverish', and 'altogether foolish and *not* boylike' (Barrett Browning and Browning 1990, 191). His solution was to turn decisively away from confessional poetry to drama: a form which allowed him to explore his psychological interests while still confirming the existence of an external world beyond the bounds of his own subjectivity. His future works would not, like *Pauline*, be 'fragment[s] of a confession' or attempts at autobiographical self-analysis. Instead, they would be psychological case studies, in which the distinction between speaker and author was rigorously maintained, and in which there would be no danger of mistaking the solipsism of his speakers as proof of the solipsism of their author.

Conclusion: The Poet's Brain

As Cook has shown, in the eighteenth century poetry had been understood primarily as a *skill*, with the great poet as one who had attained a high level of mastery in that skill through some combination of study and natural talent. It was acknowledged that such people would often be rather odd, but their oddness was not usually seen as a necessary precondition of their poetic ability (Cook 2013, 610–29). Byron and Shelley had argued that being a true poet required not only a certain kind of skill, but also a certain kind of *mind*, and in this they were followed by Horne, Hallam, Mill, and their contemporaries, who took the same argument a step further by asserting that such a mind, in turn, required a certain kind of *brain*. The poetry of Byron and Shelley had emphasised that it was necessary for the true poet to be a special kind of person, and their scandal-filled lives, eccentric lifestyles, and early deaths had served to powerfully confirm that poets really *weren't* much like ordinary people: it was for this reason that Kingsley, writing in 1853, held them to have been jointly responsible for introducing into Britain what he saw as the 'ugly superstition' of 'the right of "genius" to be "eccentric"' (Kingsley 1853, 575). But the enormously expanded definitions of madness which came into use during the period increasingly led such eccentricity to be interpreted as a form of insanity: and as a result the poet, like the criminal and the religious visionary, had by about 1830 come to be commonly seen as a mentally anomalous 'type' requiring psychological and/or physiological investigation.

It was this task that D'Israeli, Macaulay, Hallam, Mill, Horne, Madden, and Colquhoun all attempted. These seven writers all emphasised the power and value of poetry, but all save Colquhoun also drew troubling connections between poets and madmen, just as biographers and periodical reviewers had been doing throughout the 1820s (Whitehead 2017, 128). Mill and Hallam both asserted that, far from being mad, the true poet was in fact a force for sanity, capable of bringing relief to those whose psychological distress has driven them to the brink of mental collapse (Hallam 1862, 18; Gill 1998, 44–51). But their arguments for the special mental qualities of poets could easily serve as a cover for genuine insanity, as Wilson suggested was actually the case with Horne. Equipped with Horne's, Mill's, and Hallam's essays, a truly mad would-be poet could always reassure himself that the only reason everyone kept rejecting his poems as insane was because they just weren't willing to make the necessary mental effort to understand the exceptional quality of his genius.

The warnings of contemporary writers regarding the terrible toll that the writing of poetry could take on the mental health of authors must have made uncomfortable reading for aspiring young poets such as Tennyson and Browning (Madden 1833; Conolly 1830, 84, 131, 147–50, 222–3). We cannot be sure whether either Tennyson or Browning read Madden's or Conolly's books, or Croker's review of Madden in the *Quarterly*, although it is entirely possible that they did: Tennyson read widely in contemporary medical and psychological literature, and Browning could easily have learned of such ideas via his close friendship with the poet and lunacy commissioner Bryan Proctor (Faas 1988, 56–7). Even if they did not, however, all these works were indicative of a more general tendency in the writers of the period—even among the poetry lovers of the Fox circle and the Cambridge Apostles—to see poets as a mentally unstable breed, whose oddly constituted brains were constantly at risk of lapsing into madness as their faculties of judgement were overwhelmed by the power of their imaginations.

As Browning and Tennyson set out on their respective self-assigned missions to emulate the achievements of Shelley, they were confronted, much more sharply than Shelley himself had been a generation before, with the question of how the kind of visionary poetic inspiration they aspired to could be reliably distinguished from madness. Clearly, the opinion of the world at large could not be relied upon: after all, it was an article of faith for them that true poets would always be misunderstood by the general public. At the same time, however, the opinion of the would-be poet could hardly be trusted, either: for what if their conviction of their own genius was simply a symptom of their insanity? As I have shown, Tennyson and Browning explored these issues in 'Timbuctoo' and *Pauline*, respectively; but both writers continued to worry away at the epistemological problems created by apparent instances of visionary perception in the poetry that they wrote over the course of the 1830s, returning with increasing urgency to the question of how—or if—poetic vision could be distinguished from insane hallucination. It is these poems, and their relationship with the psychological controversies of the period, which form the subject of my next chapter.

References

Newspapers and Journals

The Athenaeum and Literary Chronicle. 1828–1921. London.

Primary Sources

Barrett Browning, Elizabeth, and Robert Browning. 1990. In *The Courtship Correspondence 1845–6*, ed. Daniel Karlin. Oxford: Oxford University Press.
Browning, Robert. 1983–2009. *Poetical Works*, ed. Ian Jack, Rowena Fowler, Margaret Smith, Robert Inglesfield, et al. 15 vols. Oxford: Clarendon.
Byron, George Gordon. 1921. *Poetical Works*. London: Oxford University Press.
Carlyle, Thomas. 1975. *Sartor Resartus*. London: J.M. Dent and Sons.
Colquhoun, J.C. 1836. *Isis Revelata*, 2 vols. Edinburgh: Maclachlan and Stewart.
Conolly, John. 1830. *An Inquiry Concerning the Indications of Insanity*. London: John Taylor.
Croker, John. 1834. Review of Madden's *Infirmities of Genius*. *The Quarterly Review* 50: 34–56. London: John Murray.
Crowe, Catherine. 1848. *The Night Side of Nature*. 2 vols. London: T.C. Newby.
D'Israeli, Isaac. 1822. *The Literary Character, Illustrated by the History of Men of Genius*. 3rd ed., 2 vols. London: John Murray.
Disraeli, Benjamin. 1870. *Venetia: A Novel*. New York: D. Appleton and Company.
Dryden, John. 1972. *Works*, ed. H.T. Swedenberg, Jr., 20 vols. Berkeley: University of California Press.

Gillies, Robert. 1814. *Child Alarique, a Poet's Reverie*. 2nd ed. Edinburgh: James Ballantyne.

Hallam, Arthur. 1862. In *Remains in Verse and Prose of Arthur Henry Hallam*, ed. Henry Hallam. London: John Murray.

Haydon, Benjamin. 1853. *Life of Benjamin Robert Haydon, Historical Painter, From His Autobiography and Journals*, ed. Tom Taylor, 2nd ed., 3 vols. London: Longman, Brown, Green, and Longman.

Horne, Richard Henry. 1833. *Exposition of the False Medium and Barriers Excluding Men of Genius From the Public*. London: Effingham Wilson.

———, ed. 1844. *A New Spirit of the Age*. 2 vols. London: Smith, Elder, and Co.

Kelley, Phillip, et al., eds. 1984–2016. *The Brownings' Correspondence*. 23 vols. Winfield: Wedgestone.

Kingsley, Charles. 1853. Thoughts About Shelley and Byron. In *Fraser's Magazine for Town and Country*, vol. 48, 568–576. London: John W. Parker and Son.

Lamb, Charles. 1954. *Essays of Elia and Last Essays of Elia*. London: J.M. Dent and Sons.

Landon, Letitia. 1829. *The Venetian Bracelet*. London: Longman, Rees, Orme, Brown, and Green.

———. 1839. *The Zenana, and Minor Poems of L.E.L., With a Memoir by Emma Roberts*. London: Fisher, Son, and Co.

Macaulay, Thomas Babington. 1843. *Critical and Historical Essays Contributed to the Edinburgh Review*. 2nd ed., 3 vols. London: Longman, Brown, Green, and Longmans.

Madden, Richard. 1833. *The Infirmities of Genius*. 2 vols. Philadelphia: Carey, Lee, and Blanchard.

Man Burrows, George. 1828. *Commentaries on the Causes, Forms, Symptoms, and Treatment, Moral and Medical, of Insanity*. London: Thomas and George Underwood.

Mill, John Stuart. 1981. In *Collected Works*, ed. John Robson. Toronto: University of Toronto Press.

Morison, Alexander. 1828. *Cases of Mental Disease*. Edinburgh: Maclachlan and Stewart.

Nichols, Ashton. 1986. The Epiphanic Trance Poem: Why Tennyson Is Not a Mystic. *Victorian Poetry* 24 (2): 131–148.

Reid, John. 1817. *Essays on Hypochondriacal and Other Nervous Afflictions*. Philadelphia: M. Carey.

Sandby, George. 1844. *Mesmerism and Its Opponents*. London: Longman, Brown, Green, and Longmans.

Shakespeare, William. 1984. In *A Midsummer Night's Dream*, ed. R.A. Foakes. Cambridge: Cambridge University Press.

Shelley, Percy Bysshe. 1840. *Essays, Letters from Abroad, Translations, and Fragments*, ed. Mary Shelley, 2 vols. London: Edward Moxon.

———. 1977. In *Poetry and Prose*, ed. Donald Reiman and Sharon Powers. New York: Norton.
Taylor, Isaac. 1830. *Natural History of Enthusiasm*. 4th ed. London: Holdworth and Ball.
Tennyson, Alfred. 1973. In *In Memoriam*, ed. Robert Ross. London: Norton.
———. 1982. *Letters*, ed. Cecil Lang and Edgar Shannon, Jr., 3 vols. Oxford: Clarendon.
———. 1987. *The Poems of Tennyson*, ed. Christopher Ricks, 2nd ed., 3 vols. Harlow: Longman.
Townshend, Chauncey Hare. 1854. *Mesmerism Proved True, and the Quarterly Reviewer Reviewed*. London: Thomas Bosworth.
Wilson, John. 1833. The False Medium. In *Blackwood's Edinburgh Magazine*, vol. 34, 440–468. Edinburgh: William Blackwood.
Wordsworth, William. 1992. In *Lyrical Ballads, and Other Poems, 1797–1800*, ed. James Butler and Karen Green. Ithaca: Cornell University Press.

Secondary Sources

Allard, James. 2007. *Romanticism, Medicine, and the Poet's Body*. Aldershot: Ashgate.
Armstrong, Isobel. 1996. *Victorian Poetry*. London: Routledge.
Baker, John. 2004. *Browning and Wordsworth*. Madison: Fairleigh Dickinson University Press.
Barrow, Logie. 1986. *Independent Spirits: Spiritualism and English Plebeians, 1850–1910*. London: Routledge and Kegan Paul.
Colley, Ann. 1983. *Tennyson and Madness*. Athens, GA: University of Georgia Press.
Cook, Daniel. 2013. On Genius and Authorship: Addison to Hazlitt. *Review of English Studies* 64: 610–629.
Corfield, Penelope. 1995. *Power and the Professions in Britain, 1700–1850*. London: Routledge.
Cronin, Richard. 2002. *Romantic Victorians: English Literature 1824–40*. Houndmills: Palgrave.
Day, Aidan. 2005. *Tennyson's Scepticism*. Houndmills: Palgrave Macmillan.
Faas, Ekbert. 1988. *Retreat into the Mind: Victorian Poetry and the Rise of Psychiatry*. Princeton: Princeton University Press.
Faubert, Michelle. 2009. *Rhyming Reason: The Poetry of Romantic-Era Psychologists*. London: Pickering and Chatto.
Felluga, Dino. 2005. *The Perversity of Poetry: Romantic Ideology and the Popular Male Poet of Genius*. New York: New York State University Press.

Foucault, Michel. 2006. *Psychiatric Power: Lectures at the College de France, 1973–4*, trans. Graham Burchell and ed. Jacques Lagrange. Houndmills: Palgrave Macmillan.
Fulford, Tim. 2004. Conducting the Vital Fluid: The Politics and Poetics of Mesmerism in the 1790s. *Studies in Romanticism* 43 (1): 57–78.
Gill, Stephen. 1998. *Wordsworth and the Victorians*. Oxford: Clarendon.
Herb Wright, Barbara. 1987. Tennyson, the Weird Seizures in the Princess, and Epilepsy. *Literature and Medicine* 6: 61–76.
Higgins, David. 2005. *Romantic Genius and the Literary Marketplace*. London: Routledge.
Hughes, William. 2015. *That Devil's Trick: Hypnotism and the Victorian Popular Imagination*. Manchester: Manchester University Press.
Ingram, Alan. 1992. *The Madhouse of Language: Writing and Reading Madness in the Eighteenth Century*. London: Routledge.
Jacobs, Francis. 1971. *Criminal Responsibility*. London: Weidenfield and Nicolson.
Keenan, Richard. 1973. Browning and Shelley. *Browning Institute Studies* 1: 119–145.
Kennedy, Richard, and Donald Hair. 2007. *The Dramatic Imagination of Robert Browning: A Literary Life*. Columbia: University of Missouri Press.
Kurshan, Ilana. 2006. Mind Reading: Literature in the Discourse of Early Victorian Phrenology and Medicine. In *Victorian Literary Mesmerism*, ed. Martin Willis and Catherine Wynne, 17–38. Amsterdam: Rodopi.
Lourie, Margaret. 1979. Below the Thunders of the Upper Deep: Tennyson as Romantic Revisionist. *Studies in Romanticism* 18 (1): 3–27.
Martens, Britta. 2011. *Browning, Victorian Poetics, and the Romantic Legacy: Challenging the Personal Voice*. Farnham: Ashgate.
Matthew, H.C.G., and Brian Harrison, eds. 2004. *Oxford Dictionary of National Biography*. 60 vols. Oxford: Oxford University Press.
North, Julian. 2009. *The Domestication of Genius: Biography and the Romantic Poet*. Oxford: Oxford University Press.
Pearsall, Cornelia. 2008. *Tennyson's Rapture: Transformation in the Victorian Dramatic Monologue*. New York: Oxford University Press.
Porter, Roy. 1987. *Mind Forg'd Manacles: A History of Madness in England from the Restoration to the Regency*. Cambridge, MA: Harvard University Press.
Prins, Herschel. 1995. *Offenders, Deviants, or Patients?* 2nd ed. London: Routledge.
Richardson, Alan. 2001. *British Romanticism and the Science of the Mind*. Cambridge: Cambridge University Press.
Ricks, Christopher. 1989. *Tennyson*. 2nd ed. Berkeley: University of California Press.
Rivers, Charles. 1976. *Robert Browning's Theory of the Poet, 1833–1841*. Salzburg: Universität Salzburg.

Ruston, Sharon. 2005. *Shelley and Vitality*. Basingstoke: Palgrave.

———. 2013. *Creating Romanticism: Case Studies in the Literature, Science and Medicine of the 1790s*. Basingstoke: Palgrave.

———. 2016. The Medical Dangers of Literary Genius. *Literature and Medicine* 34 (2): 299–319.

Scull, Andrew, Charlotte Mackenzie, and Nicholas Hervey. 1996. *Masters of Bedlam: The Transformation of the Mad-Doctoring Trade*. Princeton: Princeton University Press.

Shuttleworth, Sally. 1996. *Charlotte Brontë and Victorian Psychiatry*. Cambridge: Cambridge University Press.

Tate, Gregory. 2012. *The Poet's Mind: The Psychology of Victorian Poetry 1830–1870*. Oxford: Oxford University Press.

Taylor, Jenny. 1988. *In the Secret Theatre of Home: Wilkie Collins, Sensation Narrative, and Nineteenth-Century Psychology*. London: Routledge.

Tennyson, Hallam. 1897. *Alfred Lord Tennyson: A Memoir*. 2 vols. London: Macmillan.

Thomas, Donald, and Robert Browning. 1982. *A Life within Life*. New York: Viking.

Tucker, Herbert. 1988. *Tennyson and the Doom of Romanticism*. Cambridge, MA: Harvard University Press.

Whitehead, James. 2017. *Madness and the Romantic Poet: A Critical History*. Oxford: Oxford University Press.

Winter, Alison. 1998. *Mesmerized: Powers of Mind in Victorian Britain*. Chicago: University of Chicago Press.

Wood, Jane. 2001. *Passion and Pathology in Victorian Fiction*. Oxford: Oxford University Press.

CHAPTER 3

'Ah! let me not be fool'd': Delusion and Inspiration in the Poems of Browning and Tennyson, 1832–1840

This chapter explores the early poetry of Tennyson and Browning in relation to the debates over genius and madness discussed in the previous chapter. It traces their shifting attitudes towards the figure of the heroic poetic genius, and their growing scepticism regarding the Shelleyan ideal of the visionary poet-prophet. In particular, I argue that Tennyson's poems of 1832–1834 mark a crucial turning away from the visionary enthusiasms of his earlier works, while *Paracelsus* and *Sordello* dramatise Browning's movement from a 'Romantic' poetry of inspired subjectivity to a 'dramatic' poetry of objective psychological observation: a movement motivated, at least in part, by his awareness of how easily a belief in one's own inspired status could slide into a condition of solipsistic self-delusion effectively indistinguishable from madness.

Philip Van Artevelde: The Dangers of Imagination

The poetry written by Browning and Tennyson over the course of the 1830s bears witness to their ebbing faith in the authority of visionary poetic inspiration. As I have discussed, the earliest poems of both men had celebrated poets as inspired mystics, blessed with superhuman levels of insight into reality. As the decade went on, however, both Tennyson and Browning appear to have become increasingly preoccupied with the kind of false and failed visionaries described by contemporary medical writers,

© The Author(s) 2019

J. Crawford, *Inspiration and Insanity in British Poetry*,
Palgrave Studies in Literature, Science and Medicine,
https://doi.org/10.1007/978-3-030-21671-9_3

men and women who mistook fantasy, delusion, or hallucination for genuine visionary perception, usually with disastrous consequences for themselves and others.

For Browning, this shift away from the kind of sub-Shelleyan poetry that he had written as an adolescent was symbolised by his decision to destroy the unpublished second part of *Pauline*, presumably because it only offered more of the same solipsistic introversion that Mill had criticised in Part I (Martens 2011, 32; Kennedy and Hair 2007, 43). In this, Baker speculates that Browning may have been influenced by his developing friendship with Henry Taylor, whose historical 'dramatic romance' *Philip Van Artevelde* (1834) is now remembered primarily for the critique of Byron and Shelley contained in its preface (Baker 2004, 36). Of Byron and Shelley, Taylor wrote:

> It did not belong to poetry, in their apprehension, to thread the mazes of life in all its classes and under all its circumstances, common as well as romantic, and, seeing all things, to infer and instruct: on the contrary, it was to stand aloof from every thing that is plain and true; to have little concern with what is rational or wise [...] The realities of nature, and the truths which they suggest, would have seemed cold and incongruous, if suffered to mix with the strains of impassioned sentiment and glowing imagery in which they poured themselves forth. Spirit was not to be debased by any union with matter, in their effusions; dwelling, as they did, in a region of poetical sentiment which did not permit them to walk the common earth, or breathe the common air. (Taylor 1834, xii–xiii)

In this critique, the lofty sublimity of the Shelleyan or Byronic poet is turned against them, their self-declared elevation above the mundane business of life recast as a deliberate self-exile from the 'rational', the 'wise', and 'the realities of nature' which dooms them, instead, to being unwise, irrational, and out of touch with reality. But critical though he was of Byron, Taylor reserved his harshest words for Shelley:

> Mr Shelley was a person of more powerful and expansive imagination than Lord Byron, but he was inferior to him in those practical abilities, which (unacceptable as such an opinion may be to those who believe themselves to be writing under the guidance of inspiration) are essential to the production of consummate poetry. [...] Mr Shelley seems to have written under the notion that no phenomena can be perfectly poetical, until they shall have been so decomposed from their natural order and coherency, as to be

> brought before the reader in the likeness of a phantasm or a vision. A poet is, in his estimation (if I may venture to infer his principles from his practise,) purely and pre-eminently a visionary. (Taylor 1834, xxi–xxiii)

Taylor had probably not read Shelley's *Defence of Poetry*, which was not published until 1840; but if he had, he would have learned that he had been largely successful in 'infer[ring] [Shelley's] principles from his practise', even if the conclusions that he draws from them are not quite those that Shelley would have wished. Shelley *did* insist that true poets had to find new and 'vitally metaphorical' ways of describing phenomena, thus permitting them to break away from pre-existing accounts of reality in order to '[mark] the before unapprehended relations of things'; but what Shelley saw as an attempt to find a more adequate poetic language with which to describe the world appeared to Taylor as merely a deliberately fantastical refusal to present things in their 'natural order and coherency'. Similarly, Shelley *did* view the true poet as 'pre-eminently a visionary'; as he wrote in the *Defence*, the poet was one who 'beholds the future in the present' and 'participates in the eternal, the infinite, and the one'. In so doing, Shelley believed that the poet was in fact perceiving and revealing the transcendental order of the universe which lay behind the shifting surfaces of things: poets, for him, were 'those who imagine and express this indestructible order' (Shelley 1977, 482–3). But what Shelley saw as revelations of cosmic order appeared to Taylor as a state of arbitrary chaos, in which 'phenomena' are 'decomposed from their natural order' and made to collapse into unnatural 'phantasm[s]'. From Shelley's (neoplatonic) perspective, Taylor was hopelessly wedded to the transient world of appearances, and thus unable to appreciate the fact that, to the true poet, 'time and place and number are not'; but from *Taylor's* perspective, Shelley was merely an excitable young man who, because he 'believe[d] [himself] to be writing under the guidance of inspiration', had foolishly mistaken his imaginative reveries for visions of Eternity (Shelley 1977, 483).

Taylor's diagnosis of what he called 'the PHANTASTIC school' was that it showed the unfortunate results of allowing 'vision' and imagination to dominate over reason. Such works carried poetry 'to seats of anarchy and abstraction, where imagination exercises the shadow of an authority, over a people of phantoms, in a land of dreams' (Taylor 1834, xxv–xxvi). Taylor's reference to 'seats of anarchy' is probably an allusion to book 2 of *Paradise Lost*, in which Satan travels to 'the Throne / Of *Chaos*', ruler of 'Eternal *Anarchie*', but his language of 'phantoms' and 'abstractions' is very close to that used by Mill to describe the way that the mental landscape of poetry appears to the non-poet: 'a phantom-world,—a place of *ignes fatui* and spectral illusions' (Mill 1981, 1:356; Milton 1998, 407–9). However, for Taylor—who was, himself, a poet—this was not true of *all* poetry. The best and greatest poetry, in which reason is in 'equipoise' with imagination, would do nothing of the sort; instead, it would use its beauties to better enable the efficient transmission of knowledge and of moral instruction, thus granting it the kind of 'relevancy to [the reader's] life, past or future' which he saw as so grievously lacking in the works of 'the PHANTASTIC school' (Taylor 1834, xxiii–xxv).

I have discussed Taylor at some length here, not only because of his possible influence on Browning's decision to move away from the decidedly 'PHANTASTIC' poetry of *Pauline* in 1834–1835, but also because what he says here about Shelley aligns so closely with Conolly's arguments about the dangers of imagination in his 1830 *Indications*. Conolly, too, had seen imagination as a fine and valuable thing, but only as long as it was restrained by what he called 'judgement' and Taylor called 'reason'. Without this counter-balance, too much imagination could predispose the mind to insanity, which led Conolly to warn his readers that 'an irregular and injudicious cultivation of poetry and of painting has often concurred to produce madness' (Conolly 1830, 77–84, 130–1, 346). (In this, both were following the lead of Thomas Arnold, who in 1806 had warned that both 'great judgement' and 'the best moral dispositions' were necessary to prevent 'inventive genius' and 'fine imagination' from sliding into insanity [Whitehead 2017, 80].) Taylor did not actually say that Shelley was mad, but the words he used to describe his poetry—anarchy, abstraction, phantasm, vision, dream—imply a strong kinship between such undisciplined imagination and pathological mental instability.

For Conolly, the defining characteristic of insanity was not the experience of delusion or hallucination, but the failure of what he called 'the comparing faculty'. A patient was only mad if they experienced such illusions or hallucinations, *and* lacked the power to deduce, by rationally 'comparing' them to the rest of their sense-data, that the things they saw could not actually be real (Conolly 1830, 113, 300). From this perspective, it was thus essential for the would-be poet to retain some kind of contact with what Taylor called 'the realities of nature', in order to give them some kind of reference point by which their imaginative raptures might be judged. Without this, the 'comparing faculty' would have nothing to work with, and sanity could all too easily slide away.

This contact with reality was exactly what the speaker of Browning's *Pauline* had lacked. Morbidly *self*-conscious, he had been virtually unconscious of anything *outside* himself, which in the view of people like Taylor and Conolly meant that he, like the speaker of Tennyson's 'Timbuctoo', might be dangerously at risk of mistaking the 'angel' of his imagination for an actual angel (Browning 1983–2009, 1:44). As I have discussed, Browning's solution to this problem was to turn to dramatic poetry, and the first result of his dramatic labours was *Paracelsus* (1835), a closet drama which suggests the extent to which he had taken Taylor's warnings to heart.

Paracelsus: The Failure of Pure Imagination

Browning's *Paracelsus* stages in dramatic form the very opposition which Hallam, Taylor, Mill, and Conolly had all identified as existing between the visionary insights of Shelleyan poetry and the rational, practical knowledge which governs the external world: an opposition which had led Conolly to recommended that lovers of poetry should also study mathematics, in the hope that the stabilising influence of the latter would balance out the dangerous effects of the former (Conolly 1830, 192–3, 346–7). In the second part of *Paracelsus*, Browning depicted this opposition between practical reason and visionary imagination in extremely literal terms, in a scene in which his eponymous protagonist—who has devoted his life to scientific knowledge—encounters a dying poet, Aprile, who is a kind of composite of Shelley and Keats (Keenan 1973, 126). (He also sounds a great deal like the speaker of *Pauline*; indeed, Aprile's death-scene in *Paracelsus* may be read as Browning killing off 'the *Poet* of the batch' whom he had intended to masquerade as two years before.) As Paracelsus swiftly comes to realise, he and Aprile are equals and opposites, 'halves of one dissevered world', so much alike that Paracelsus initially assumes that Aprile must be a rival scholar, while Aprile mistakes Paracelsus for a fellow-poet (Browning 1983–2009, 1:236–40).

Where Paracelsus has sought only knowledge and mastery, organising his understanding of the world in terms of practical reason, Aprile has sought only beauty and love, organising *his* world along purely aesthetic and emotional lines, and thus demonstrating himself to be precisely the kind of 'born poet' whom Mill and Hallam had praised. Paracelsus soon concludes that, in adopting such one-sided approaches to life, both he and Aprile have been badly in the wrong, and he begs the dying poet to help him change his ways:

> Love me henceforth Aprile, while *I* learn
> To love—& merciful God forgive us both!
> We wake at length from weary dreams. but both
> Have slept in fairy-land—tho' dark & drear
> Appears the world before us, we no less
> Wake with our wrists & ancles jewelled still:
> I, too, have sought to KNOW as thou to LOVE
> Excluding love as thou refused'st knowledge:
> Still thou hast beauty & I power: we wake:
> What penance canst devise for both of us? (Browning 1983–2009, 1:262)

In Conolly's terms, this is Judgement addressing Imagination. Each has achieved something of real value—'power' and 'beauty', respectively—but in trying to live without each other, both have become estranged from reality, lost in 'weary dreams'. As result, both have come to grief: Paracelsus has been reduced to despair, and Aprile had been brought to the point of death. Paracelsus realises that each of them has the power to save the other, and begs Aprile not to leave him 'Till thou, the Lover, know,—& I, the Knower, / Love—until both are saved!' (Browning 1983–2009, 1:264). But before this happy unity between love and knowledge, reason and sensation, judgement and imagination, can be realised, Aprile dies in his arms; and Paracelsus, his lesson learned, moves on to the next stage of his life.

Aprile is a failed visionary, and he knows it. Indeed, it seems to be his awareness of his own failure which is killing him, just as it may have already driven him mad. Like the speaker of *Pauline*, Aprile is tormented by the belief that he can never live up to the example of the great poets whom he loves, and when Paracelsus first meets him he is already lamenting his own failure. His first words in the poem are these:

> Lost one, come! the last
> Who, living, hast life o'erpast
> And, all together we
> Will ask for us & ask for thee
> Whose trial is past whose lot is cast
> With those who watch but work no more,—
> Who gaze on Life, but live no more:
> Yet we chose thee a birth-place
> Where the richness ran to flowers …
> Could'st not sing one song for us? (Browning 1983–2009, 1:232)

These are very enigmatic verses, not least because they leave deeply unclear who exactly 'we' are supposed to be. In his 1849 revision of *Paracelsus* Browning expanded this part of the poem in order to clarify what, exactly, is going through Aprile's mind, and adding an introductory passage in which Aprile describes how he can hear the voices of dead poets singing to him out of the air:

> As if all poets, God ever meant
> Should save the world, and therefore lent
> Great gifts to, but who, proud, refused
> To do his work, or lightly used
> Those gifts, or failed through weak endeavour,
> So, mourn cast off by him for ever,—
> As if these leaned in airy ring
> To take me; this the song they sing. (Browning 1983–2009, 1:231–3)

Like Shelley, Aprile has an exalted opinion of the power of poetry, asserting that the reason that God gives 'Great gifts' to poets is so that they can 'save the world', a task that one might have thought was reserved for the messiah. However, while Shelley saw himself as part of a tradition of successful visionary poets stretching back through history to Milton, Dante, and Homer, Aprile takes a more melancholy view of his historical position: he is the latest in a long line of poets who have failed to accomplish the sacred task their poetic gifts were meant to equip them for, and who have accordingly been 'cast off' by God.

In the expanded 1849 version of his opening song from the 1835 text, Aprile hears or imagines a chorus of dead, failed poets, commiserating with him in his despair:

> So sank we in those old years,
> We who bid thee, come! thou last
> Who, living yet, hast life o'erpast.
> And altogether we, thy peers,
> Will pardon crave for thee, the last
> Whose trial is done, whose lot is cast
> With those who watch but work no more,
> Who gaze on life but live no more.
> Yet we trusted thou shouldst speak
> The message which our lips, too weak,
> Refused to utter,—shouldst redeem

Our fault: such trust, and all a dream!
Yet we chose thee a birthplace
Where the richness ran to flowers:
Couldst not sing one song for grace?
Not make one blossom man's and ours?
Must one more recreant to his race
Die with unexerted powers,
And join us, leaving as he found
The world, he was to loosen, bound? (Browning 1983–2009, 1:233–5)

Aprile is crushingly aware of his own failure to live up to the (impossibly high) standards he has set for himself, and for the art of poetry in general. Because he has failed to save the world with his poems, he is a 'recreant to his race', a disappointment both to God and to the spirits who have watched over him since 'first / The waters of my life outburst', and having thus failed them he has nothing left to do but die.

Aprile, then, offers an example of a Shelleyan quest for visionary poetic power terminating in disaster rather than triumph; and as Taylor and Conolly would have predicted, his failure is due to his inability to bring order to his unruly imagination. As a result it has quite overmastered his judgement, as he explains to Paracelsus in the 1835 version of the poem:

Didst thou, when nerving thee to this attempt,
Ne'er range thy mind's extent, as some wide Hall,
Dazzled by shapes that filled its length with light,
Shapes clustered there, to rule thee, not obey,—
That will not wait thy summons—will not rise
Singly,—nor when thy practised eye & hand
Can well transfer their loveliness,—but crowd
By thee for ever, bright to thy despair! (Browning 1983–2009, 1:258)

This is the ever-present 'angel' imagination of *Pauline* reimagined as a curse rather than a blessing. Rather than making his creative imagination a light to see by, Aprile has been 'dazzled' by it, and, as a result, the fact that it is always present has become intrusive and distracting rather than comforting. The creations of his imagination are 'there to rule [him], not obey': they overwhelm his reason, refusing to 'wait [his] summons', and crowd into his mind regardless of whether he wants to or not.

The very fact that Aprile experiences his imaginative life in this way, as a chaotic mass of 'shapes' over which he has no conscious control, demonstrate the extent to which he has failed to achieve that level of self-mastery which, for Taylor and Conolly, was the mark of the truly great poet. As Wood notes, it was widely believed at the time that 'imagination ... required constant regulation by a controlling will if the stigmatising symptoms of nervous illness were to be kept at bay' (Wood 2001, 60). Aprile, however, is as utterly overwhelmed by the creations of his imagination as the madman is by his delusions and hallucinations, and is thus unable to produce the great work for which his gifts should have equipped him. He has a 'practised eye & hand', but rather than waiting patiently for the moment when he is best able to use them to 'transfer their loveliness' into successful art, the 'shapes' of his imagination simply mob him at every opportunity. As a result, rather than fulfilling his artistic ambitions, he has suffered exactly the consequences which Madden would have expected from such indiscriminate artistic and imaginative activity, and has declined into mental instability, despair, and early death. Being thus enslaved by his own imagination, it is hardly surprising that Aprile is initially dismissed by Paracelsus as a harmless lunatic, merely 'some moonstruck creature after all!' (Browning 1983–2009, 1:242).

Browning clearly did not wish to be too harsh on Aprile, perhaps because he rather identified with him; his perspective, after all, is very similar to that which Browning himself had adopted just a few years before. As Aprile dies, he has a vision of poets—not failed poets, this time, but successful ones, with 'White brows—lit up with glory'—who come to crown him, despite his protestations that they have come to the wrong man, implying that his failure as a poet has not been nearly as great as he himself believed (Browning 1983–2009, 1:264). As Paracelsus recognises, Aprile has not been wrong to devote his life to love and imagination; he has gone astray only in his failure to combine imagination with reason, and love with knowledge. But even if Aprile has been a truer poet than he himself knows, he has manifestly not achieved his poetic goals; and even after having recognised him as his equal and opposite, Paracelsus continues to refer to Aprile as 'the poor mad poet' (Browning 1983–2009, 1:302).

Taken as a whole, the 1835 *Paracelsus* thus offers up a very qualified endorsement of the kind of Shelleyan poetry that Mill had praised in 1833 and Taylor had condemned in 1834. Whereas Taylor had seen imagination without reason as basically worthless, and Conolly had seen imagination without judgement as positively dangerous, Browning presents it as something of real value—but as only *part* of the truth. Paracelsus reveres Aprile's memory, but he also recognises that 'the poor mad poet' was wrong on two important counts: imagination without self-control and love without knowledge are ***not*** sufficient to work the world's salvation, and in any case the 'poet who shall save the world' which Aprile reproached himself for not becoming is an impossibility, because the world must be redeemed gradually over the course of ages rather than being saved all at once in a single moment of Shelleyan revolution. When Paracelsus lies dying at the end of the play, his last words are 'Aprile! Hand in hand with you, Aprile!'—partly because he still loves and admires the poet, but also because he recognises that it is only when they are combined, 'hand in hand', that they add up to something truly worth having:

> [L]et Men
> Regard me, & the Poet dead long ago
> Who loved too rashly—& shape forth a Third
> And better tempered spirit warned by both,
> As from the over-radiant star too mad
> To drink the light-springs, beamless thence itself—
> And the dark orb which borders the abyss,
> Ingulphed in icy night,—might have its course
> A temperate & equidistant World[.] (Browning 1983–2009, 1:496–8)

The 'over-radiant star too mad / To drink the light-springs, beamless thence itself' stands, of course, for visionary poets such as Aprile, who are thus once again associated (as they were by Madden) with madness, and with a radiance more likely to cause burnout and blindness than illumination. (In Browning's 1849 rewrite of *Paracelsus*, the phrase 'too mad / To drink the light-springs' was changed to 'too mad / To drink the life-springs', further cementing the link between 'over-radiant' poetic creativity, insanity, and early death.) Such poetry, the play suggests, is a dangerous pursuit, likely to exact a very heavy toll on the mind and body of the poet, and offering neither true knowledge nor true power in exchange. Much wiser, then, to take a 'temperate and equidistant' course, combining love and knowledge, reason and imagination, beauty and power and truth; much wiser to steer away from the more extreme forms of visionary poetry, and to shun the imaginative excesses which had led Aprile and his real-world equivalents to their early graves.

For the poet, then, just as much as for the victim of hallucinations, it was essential to cultivate what Conolly had called 'the comparing faculty'. As Rivers argues, in *Paracelsus* Browning 'subordinates intuition to knowledge' (Rivers 1976, 86). Left to its own devices, the pure imagination which Taylor saw as characterising the poetry of 'the PHANTASTIC school' would rapidly carry the poet out of touch with reality, and if their imaginings, feelings, and sensations were not tested and measured against some kind of firm knowledge of the external world, they could all too easily spiral away into solipsism and insanity. Browning's subsequent poetry, from 'Porphyria's Lover' (1836) all the way to *Red Cotton Night-Cap Country* (1873), frequently concerned itself with characters of precisely this kind: individuals who have allowed their own private imaginations and delusions to carry them out of touch with reality, usually to the detriment of themselves and others. Unlike *Pauline*, however, these poems made clear that their protagonists were not to be identified with their author, and it was through this implicit contrast between the characters (who have mistaken imagination for reality) and their author (who has not) that Browning was able to demonstrate that his own 'comparing faculty' remained very much intact.

Faith and Doubt: 'St Simeon Stylites'

Like Browning, Tennyson read Taylor's critique of 'the PHANTASTIC school', and while Tennyson argued that Taylor '[did] not take sufficiently into consideration the *peculiar* strength evolved by such writers as Byron and Shelley', he also admitted that the preface 'makes me shamed of my own faults' (Tennyson 1982, 1:120). By 1834, however, his own poetry had already moved some distance from the Shelleyan faith in visionary inspiration that had characterised his works in 1828–1830, and towards poems which repeatedly addressed the question of how to reliably distinguish between visions and delusions, appearances and realities. If, for example, one has a vision of an angel, like those experienced by the poet in 'The Mystic' or the speaker in 'Armageddon', it will obviously be tempting to take it at face value, not least because if what you have seen is a real angel, then you may be a real prophet. But might it not be safer to assume, like the modern, educated ghost-seers described by Scott's *Letters on Demonology*, that one is very much more likely just to be experiencing some kind of natural coincidence or unusual mental episode instead? Should visionary experiences be regarded as encounters with a higher reality, 'more real' than ordinary objects of sense perception, or as mere delusions, 'less real' than the everyday objects around them?

In Tennyson's 'St Simeon Stylites', this is the cruel dilemma which confronts Simeon upon his pillar, where after 30 years of ascetic self-mortification he is desperate for some confirmation that his sufferings have earned him a place in paradise:

> While I spake then, a sting of shrewdest pain
> Ran shrivelling thro' me, and a cloudlike change,
> In passing, with a grosser film made thick
> These heavy, horny eyes. The end! the end!
> Surely the end! What's here? a shape, a shade,
> A flash of light. Is that the angel there
> That holds a crown? Come, blessèd brother, come,
> I know thy glittering face. I waited long;
> My brows are ready. What! deny it now?
> Nay, draw, draw, draw nigh. So I clutch it. Christ!
> 'Tis gone: 'tis here again; the crown! the crown!
> So now 'tis fitted on and grows to me,
> And from it melt the dews of Paradise,
> Sweet! sweet! spikenard, and balm, and frankincense.
> Ah! let me not be fool'd, sweet saints: I trust
> That I am whole, and clean, and meet for Heaven. (Tennyson 1987, 1:603–4)

'What's here?' That, indeed, is the question; and the line ''Tis gone: 'tis here again' alludes to another famous literary apparition, Old Hamlet's ghost (''Tis here / 'Tis here / 'Tis gone'), whose ontological status is similarly questionable. Simeon sees 'a shape, a shade / A flash of light', which sounds ambiguous enough that it could be anything, but he leaps to put a favourable interpretation upon it: 'Is that the angel there / That holds a crown?' Well, is it? Simeon wants it to be, and he feels that he *deserves* it to be, and he hopes that this makes it logical for him to conclude that this is what it actually is, but his anxiety remains acute. 'Ah! let me not be fool'd, sweet saints', he begs, still painfully aware that this could all just be delusion.

Like the poet-prophets described in Tennyson's earlier poems, Tennyson's Simeon proclaims himself to be in communication with supernatural forces. Unlike them, however, he is wracked with doubt, horribly uncertain of whether he can trust the visions which have resulted from his ordeals. Indeed, he has reduced himself to such a state of bewilderment that he is no longer sure of anything, even the penances of which he boasts:

I think that I have borne as much as this—
Or else I dream—and for so long a time,
If I may measure time by yon slow light,
And this high dial, which my sorrow crowns—
So much—even so.
              And yet I know not well,
For that the evil ones come here, and say,
'Fall down, O Simeon: thou hast suffered long
For ages and for ages!' then they prate
Of penances I cannot have gone through,
Perplexing me with lies; and oft I fall,
Maybe for months, in such blind lethargies,
That Heaven, and Earth, and Time are choked. (Tennyson 1987, 1:598–9)

'And yet I know not well'; there is a world of pain in those six words. Simeon thinks he has been on his pillar for 20 years, and engaged in his ascetic self-mortifications for 30, but he cannot be sure: perhaps he dreamed it, or perhaps it is one of the lies with which 'the evil ones' 'perplex' him. (And just what *are* these 'evil ones'? Jeering onlookers? Demonic spirits? Possibly Simeon himself is no longer sure.) His body and mind are disintegrating, and he can no longer reliably distinguish between reality, hallucination, and dream.

Simeon reels off a long list of appalling 'penances' which he claims to have inflicted upon himself, but it is not clear whether his account is at all reliable: how do we (or he) know that these are not just more of the 'penances I cannot have gone through', of which the lying 'evil ones' love to 'prate'? (Even worse is the possibility that they aren't lying at all, and his actual sufferings have been even more extreme—so terrible, in fact, that he refuses to acknowledge them as true, insisting that he 'cannot have gone through' them and that his memories of them must be demonic 'lies'.) He repeatedly boasts of the duration of his penances, but by his own account he can no longer keep track of time, 'oft' falling into states of total physical and mental collapse in which 'Time [is] choked'; he guesses that these last 'maybe for months', but it sounds as though he has no real way of knowing whether it is days or years. (A few months later, Tennyson would write in 'The Two Voices' of how the memories of madmen 'Oft lose whole years of darker mind') (Tennyson 1987, 1:590). Any or all of his story could be 'dream' or hallucination: as Platizky notes, even the 'dews of Paradise' he boasts of 'may be nothing more than feverish sweat' (Platizky 1989, 34–5). All that he can really be sure of is that he is suffering intensely, that he thinks he has been suffering for a very long time, and that he desperately hopes that his suffering has not been in vain.

What 'St Simeon Stylites' so grotesquely illustrates are the difficulties which must be faced by anyone who consigns themselves to a potentially psychologically damaging way of life—by devoting oneself to writing poetry, for example—in the hope of attaining visionary insight. Even if it seems to work, how can one ever trust the resulting visions, knowing how much physical and psychic damage one is likely to have sustained in pursuit of them? As Cronin notes of 'St Simeon' and Browning's near-contemporary 'Johannes Agricola', 'it is possible to read these, the very first Victorian monologues, as a mordant response to the belief so closely associated with Wordsworth that the poet's true function was as a dispenser of religious truths' (Cronin 2012, 47). Tennyson's Simeon is, in some ways, a ghastly parody of the inspired Romantic artist, and a worrying amount of what Tennyson said about the ideal poet in 'The Mystic', 'The Poet', and 'The Poet's Mind' could be applied with equal accuracy to poor St Simeon. Like Wordsworth on Snowden, or Byron's Manfred, Simeon has physically raised himself above the level of ordinary life in order to commune with spiritual realities. Like Blake or Shelley, he has attempted through supreme effort to transcend the commonplace world of things and gain access to the truer, higher world beyond it. But like Tennyson, and unlike the self-proclaimed poet-prophets of the previous two generations, Simeon remains unsure of whether his visions are genuine.

Even when Simeon does make claims for his own special status, the poem hardly encourages its readers to take his assertions at face value. There were so many other possible explanations which could account for his experiences: and the works of the new psychologists were full of confident explications of cases such as Simeon's, sympathetically demonstrating that such delusions were simply the result of illness, or enthusiasm, or wrong association, or imagination overpowering judgement, or physical malfunctions in the brain. Morison, for example, records the case of one man who thought 'that holes in the walls are full of spirits', and believed himself to have been 'favoured by God with superior wisdom'. Morison claimed to have cured him of both beliefs with the aid of a three-month regime of calomel (Morison 1828, 72–3).

When Tennyson wrote 'St Simeon Stylites' in 1833, he was at a pivotal moment in his life. In 1831, after his father had finally succeeded in drinking himself to death, Tennyson had reassured his anxious uncle that he would, like his late father, take up a career in the church, thus making some practical use of his expensive university education (Batchelor 2014,

57). But when he finally left Cambridge in 1832, he showed no intention of joining the clergy; instead, he decided to devote himself entirely to poetry. In one sense, 'St Simeon Stylites' may have been about the church career which his family had forced upon his father, and were now trying to force upon him: service to God imagined as a kind of slow-motion martyrdom, with Simeon's 30 years of penance corresponding to the 30 years that George Tennyson had spent as a clergyman, from his ordainment in 1801 to his death in 1831. Tennyson's Simeon's resembles the poet's father in other ways, as well: his 'blind lethargies' sound like George Tennyson's alcoholic stupors, while Simeon's difficulty in differentiating between fact and fiction recall his tendency to invent fantastical stories about his own adventures (Ricks 1989, 23–4). The poem is framed as a death-bed monologue—Simeon himself certainly believes he is dying as he speaks—and when Tennyson wrote it the delirium of his dying father, who had 'talk[ed] wildly and [was] evidently deranged' in the days leading up to his death, may not have been far from his mind (Batchelor 2014, 57). The poem imagines what it might be like to look back across a life sacrificed to the service of an abstract ideal, and how the desperation to believe that such a sacrifice had been worthwhile could drive someone into a state indistinguishable from madness.

'St Simeon Stylites', then, can be seen as an expression of Tennyson's scepticism regarding the value of devoting his life to the church. But it may also be expressing another anxiety: the worry of what his life might be like if he got his wish and devoted himself to poetry, instead. For if he, like Simeon, rejected normal life in order to pursue something higher, sacrificing the relative comfort and security which could be his in favour of an uncertain future as a poet—a profession which, as I have discussed, was notorious at the time for the mental strains which it imposed upon its practitioners—then he, too, might end up attaining a state of visionary perception only after his mind had been so damaged that he could no longer reliably discern which of his visions were false and which were true.

Epistemological Despair: Tennyson's Poems of 1832–1833

'St Simeon Stylites' was written at a time during which Tennyson was haunted by the fear of descending into madness (Colley 1983, 51–5; Batchelor 2014, 56). His own state of mind was not encouraging, and in the summer of 1832 Hallam was alarmed by Tennyson's 'overanxious thought' and 'morbidly intense ... inward concentration': the same symptoms which led, a few months later, to his brother Edward's permanent confinement (Tennyson 1982, 1:60, 1:68). But much worse was to come: and Hallam's death by cerebral haemorrhage in 1833, itself a reminder of the apparent weakness of poetic brains, plunged Tennyson into a deep depression in which life seemed meaningless and suicide dangerously inviting (Faas 1988, 53–5; Colley 1983, 52).

Tennyson's feelings of 'depression' and 'bitterness of spirit' during this period are vividly evoked in 'The Two Voices', a poem which he began writing before the news of Hallam's death arrived, but which he afterwards came to associate with the period of grief which followed it, during which the poem was completed (Tennyson 1982, 1:97, 1:105). In it, he describes the voice of despair:

> Again the voice spake unto me:
> "Thou art so steep'd in misery,
> Surely 'twere better not to be.
>
> "Thine anguish will not let thee sleep,
> Nor any train of reason keep:
> Thou canst not think, but thou wilt weep." (Tennyson 1987, 1:573)

As I have mentioned, Coleridge had argued that mental experience was basically unitary, and that this disproved Gall's hypothesis that the brain was a mass of competing 'organs' (Richardson 2001, 48–9). In 'The Two Voices', however, Tennyson describes a mind divided against itself, fissured into multiple disputing 'voices' by the weight of the 'anguish' that it has experienced. As in 'Timbuctoo' and 'St Simeon Stylites', the resulting mental experience disguises itself in the trappings of prophetic revelation, this time in the form of a 'still small voice' which speaks to him like the voice of God which Elijah heard upon Mount Carmel. Instead of bringing him divine comfort, however, this 'still small voice' brings the speaker a message of suicidal despair. In the place of unified vision comes self-division, the speaker's psyche splitting into a desperately arguing 'I' and a voice that sounds like God but is almost certainly just a symptom of the speaker's 'crazy sorrow', and which mocks him for lacking 'A healthy frame' and 'a quiet mind', dismissing his attempts to dispute with it by telling him: 'Sick art thou, a divided will' (Tennyson 1987, 1:576, 1:591). When confronted with seemingly supernatural events (such as an apparently-divine voice that tells you to kill yourself), Scott, Hibbert, Conolly, and Thacher had argued that it was essential to apply rational judgement in order to determine whether such experiences were not more likely to be mere hallucinations. But in 'The Two Voices', this strategy proves doubly ineffectual: firstly because the speaker's mentally shattered state makes it impossible for him to maintain any such 'train of reason', and secondly because the 'still small voice' pre-empts any attempt to discount it as a mere pathological symptom by arguing that if the speaker is so sick that he's hearing voices, then that's just another reason why he should surrender to despair. Under such circumstances, it is hardly surprising that all he can do is weep.

Attempting to silence the voice's corrosive scepticism, the speaker, in a very Wordsworthian fashion, cites his own mystical experiences as proof that his life must have some kind of meaning and value:

> Moreover, something is or seems,
> That touches me with mystic gleams,
> Like glimpses of forgotten dreams—
>
> "Of something felt, like something here;
> Of something done, I know not where;
> Such as no language may declare." (Tennyson 1987, 1:590)

In a more traditional poem of imagination impaired and restored, such as 'Tintern Abbey', these 'mystic gleams' would be the antidote to the speaker's despair, just as they were for Wordsworth: his thoughts *tell* him that life is meaningless, but his feelings *show* him that this is not the case. In 'The Two Voices', however, they prove just as ineffectual as rational argument, largely because the 'gleams' themselves seem to be far too vague and doubtful to serve as the foundation of any kind of Wordsworthian certainty. By describing his mystical experience as something that 'is or *seems*' (my italics), rather than as something that definitively *is*, the speaker gives the game away: just because he feels these things does not mean that they possess any more reality outside his own imagination than does the god of the priestess in 'Timbuctoo'. The voice of 'crazy sorrow' in his mind gives short shrift to such visionary vagueness:

> The still voice laugh'd. "I talk," said he,
> "Not with thy dreams. Suffice it thee
> Thy pain is a reality." (Tennyson 1987, 1:590)

The speaker's mystic pretensions crumble before the unquestionable subjective reality of his mental distress. What ultimately saves him from despair is not any of the rational arguments he attempts to proffer, still less his 'mystic gleams', but the appearance within his mind of yet *another* voice that promises him 'a hidden hope', and which he chooses to listen to instead.

'The Two Voices' invokes two earlier poems by Coleridge, 'Dejection: An Ode' and 'The Rime of the Ancient Mariner', which enact similar voyages through despair into hope and spiritual regeneration. The voice of despair vanishes when the speaker blesses the family walking to church, just as the albatross falls from the neck of Coleridge's ancient mariner when he blesses the water-snakes, and just as Coleridge's 'Dejection' ends with new hope and the words 'ever, evermore rejoice', the last words of 'The Two Voices' are 'Rejoice! Rejoice!' (Tennyson 1987, 1:593; Coleridge 1999, 399). Yet while the poem presents the speaker's decision to heed the second voice rather than the first as a triumph of hope and faith over melancholy, this is largely because Tennyson chooses to break off at a moment when the second voice is uppermost, and it is by no means clear that the poem as a whole represents anything more than a single manic-depressive cycle within a longer episode of acute mental distress. For the majority of the poem, the 'Two Voices' of the title seem to be the rational voice of the speaker, and the sceptical voice of despair he argues with. By the end of the poem, however, it turns out that there are in fact at least *three* voices in the speaker's mind, raising the possibility that what the poem describes is not a process of spiritual reintegration but one of progressive mental disintegration into mutually antagonistic warring parts—the very opposite of the unified mind of the poet which Coleridge had described. As Tate observes, 'The title of the poem sets up a binary, but the fragmentation of the mind goes beyond this: there are three voices in "The Two Voices" and the poem examines a mind that, even as it changes, remains permanently and essentially divided' (Tate 2012, 49). Certainly Tennyson himself continued to experience feelings of despair and depression for some years after this poem was written, despite its optimistic conclusion.

These years of weeping were also those in which Tennyson's elder brother Charles was retreating into opium addiction, perhaps in order to console himself for the fact that he, unlike Tennyson, actually *had* followed their father into a church career in 1833. (In this, too, he was following in his father's footsteps: George Tennyson had made heavy use of opium in the last years of his life.) (Tennyson and Dyson 1974, 64; Stevenson 1982, 119–22; Platizky 2002, 209). That Tennyson himself was acutely aware of the temptations of such narcotic oblivion is obvious from his poem 'The Lotos-Eaters', written between 1830 and 1832, in which Odysseus' sailors make a passionate plea to be allowed to retreat from the troubles of life into a haze of drug-induced bliss which, as Stevenson notes, resembles the dream-states praised by De Quincey, Coleridge, and Keats (Stevenson 1982, 126–31). Their longing for narcotised sleep is not difficult to understand: it brings an end to all distress and pain. But for someone as epistemologically anxious as Tennyson, deliberate surrender to the 'dreamful ease' of opium-reverie offered an additional temptation: the abdication of epistemological responsibility. Simeon, like the speakers of 'Timbuctoo' and 'The Two Voices', worries over whether his experiences are illusions or realities, but the lotos-eaters knowingly embrace illusion, indifferent to whether or not the things they see or feel are real so long as they bring them pleasure. Had Simeon been a lotos-eater, he wouldn't have cared whether the angel he thought he saw was real or not. He would just have been grateful that its appearance, whether real or illusionary, served to bring some comfort to his dying hours.

The same dilemma with which the lotos-eaters confront Odysseus—is a life of real suffering actually preferable to a life of illusionary comfort?—is posed in reverse by another of Tennyson's poems from this period, 'The Lady of Shalott'. The Lady knows that she is living in a world of 'shadows' rather than realities, and that, like a far-gone opium addict, she is unlikely to survive the shock of trying to live in any other way. Like the lotos-eaters, she is set apart from the world outside, suspended in enchanted isolation from all the ordinary cares of life: she 'weaveth steadily / And little other care has she'. Unlike them, however, she does not view this as a blessing. The lotos-eaters retreat into their shadow-life as a refuge from a lifetime of suffering; but the Lady seems to have been born into it, and has lived amongst shadows until she is 'half sick' of them (Tennyson 1987, 1:390–1).

In 'Timbuctoo', the Spirit of Fable asserts that the role of fables is to teach us 'to attain / By shadowing forth the Unattainable'. He thus implies a threefold model of reality: there is the unreal (but still useful) shadow-world of Fable, the real world in which we strive to 'attain' real objectives, and finally the numinous realm of 'the Unattainable', which can be 'shadow[ed] forth' but presumably not directly experienced. (It would hardly be 'Unattainable' otherwise.) The most obvious reading of the Lady's situation places her in the world of shadows, yearning towards the world of real attainment: she is a person who has been forced to live amongst representations rather than realities, until finally she resolves to seek the real even if it kills her. But what if her yearning is for 'the Unattainable', instead? The Lady's imprisonment can function just as easily as a metaphor for consciousness itself: if her tower is a body and her room is a skull, then her window is an eye-socket, her mirror is a retina, her loom is a brain, and the Lady herself is the animating spirit or soul. In this case, the shadow-world that she is half sick of would be the world of ordinary perception, of reality as filtered through our limited, fallible senses and interpreted by our limited, fallible brains. She would be like one of the prisoners in Plato's cave, as Shannon points out, and her true longing would be for an unmediated, visionary glimpse of true reality (Shannon 1981, 211). She would, in other words, be an aspiring mystic—or a Romantic poet.

There is some textual support for such a reading. When she leaves the tower, the Lady is described as being 'Like some bold seër in a trance', and as she sails, dying, along the river to Camelot, she becomes, however briefly, a kind of inspired bardic figure, singing the 'longdrawn carol, mournful, holy' which will be her 'deathsong'. (In the 1832 version of the poem, she even achieves the quintessentially Romantic feat of not being understood by 'the wellfed wits' who gather around her body.) But the price is appallingly high. Her vision blinds her: as she sings, 'her eyes [are] darkened wholly', and she dies in mid-performance, not even having had a chance to finish it (Tennyson 1987, 1:393–4). Noting the Lady's resemblance to Shelley's Alastor, Lourie writes: 'Oddly, Tennyson here adopts Shelley's metaphor of inner exploration for the voyage out.' But this is only 'odd' if the Lady is understood as 'voyaging outward into the physical world', rather than as someone making an Alastor-esque attempt to break through the Platonic shadow-world of appearances and reach some higher reality beyond it (Lourie 1979, 19–20). However, like Alastor—and, one might add, like Simeon—the Lady discovers that the price of vision is the destruction of the visionary.

Collectively, Tennyson's poems of 1832–1833 thus mark a striking departure from the visionary optimism of 1828–1830. In place of 'The Poet' and 'The Mystic', completely certain that they had tapped into the deep truth behind the universe and eager to share it with other, less-enlightened souls, we have the Lady (whose search for a more adequate knowledge of reality kills her before she is able to successfully communicate her experiences to others), the Lotos-Eaters (who have been so exhausted by their striving that they conclude, in despair, that illusions are preferable to reality), St Simeon (who is so mentally damaged that he can no longer be sure whether his visions are imaginary or real), and the speaker of 'The Two Voices' (who, when he hears something which sounds like the voice of God, discovers that all it has to say to him is that he is sick and mad, that his mystical experiences are meaningless dreams, and that he should kill himself as soon as possible). All, in their own ways, are failed Romantics, perhaps even parodies of Romanticism: the Lady a failed Keats who dies so romantically young that she never finishes a single work, the Lotos-Eaters a gang of failed Coleridges for whom opium dreams have replaced real achievements, St Simeon a failed Wordsworthian recluse whose separation from ordinary life has brought him pain and bewilderment rather than wisdom and serenity. Indeed, St Simeon and the Lotos-Eaters resemble these poets as they were sometimes described by their less sympathetic critics: men who have sunk into solipsism, mistaking their own speculations and fantasies for divine, universal truths. Tellingly, after 1830, 'The Mystic' was not reprinted in any of Tennyson's subsequent poetry collections (Colville 1970, 193).

Tennyson's willingness to leave his various sufferers at their lowest ebb—to leave the lotos-eaters chewing their lotos and Mariana longing for her death, even though in both cases they were drawn from larger stories in which this was not their ultimate fate—was noted by Croker in his characteristically vicious review of Tennyson's *Poems*, by which Tennyson was deeply hurt:

> How they got home you must read in Homer:—Mr. Tennyson—himself, we presume, a dreamy lotus-eater, a delicious lotus-eater—leaves them in full song.
>
> Next comes another class of poem—Visions. The first is the "Palace of Art", or a fine house, in which the poet *dreams* that he sees a very fine collection of well-known pictures. An ordinary versifier would, no doubt, have followed the old routine, and dully described himself as walking into the Louvre, or Buckingham Palace, and there seeing certain masterpieces of painting: a true poet dreams it. (Jump 1967, 78–9)

By describing Tennyson as a 'lotus-eater' just before going on to discuss his dream-poems, Croker hints strongly that he viewed Tennyson himself as sharing the faults of the lotos-eating mariners whom he described: that he was someone who, like them, had wilfully chosen to live in a world of dreams and visions rather than engaging with the real, to the point of actively preferring fantasies to realities. He could have gone out into the world and had a real experience, seeing real paintings in a real building such as 'the Louvre, or Buckingham Palace', but, instead, he prefers to simply dream about doing so, apparently in the belief that this exaltation of dream above reality will mark him out as 'a true poet'. This, Croker implies, is why Tennyson leaves his lotos-eaters 'in full song' rather than narrating their departure from the island; being himself 'a dreamy lotus-eater' he has no interest in real life, only in narcotic hallucinations, visions, and dreams. This is rather unfair to Tennyson—being unfair to Romantic and would-be Romantic poets was Croker's stock-in-trade—but it *was* true was that Tennyson, having imagined his various dream- or nightmare-states, often seemed to have trouble imagining his way back out of them, at least in ways that did not end with death. As Tucker has noted, 'Tennyson's imagination was reluctant to believe in the efficacy of human action', and in his works 'instances of independence are rare'. More commonly, his protagonists remain paralysed, able to act only as a *reaction* to what Tucker calls 'some distant final cause' (Tucker 1988, 17).

Assailed by deep depression, and anxious about his odd trance states and his family history of epilepsy, Tennyson worried that he risked falling into mental states from which his own mental resources might be insufficient to deliver him, especially in light of contemporary warnings over the psychological dangers posed by over-indulgence in poetry. While he directly addressed his despair, and his hopes for future recovery, in 'The Two Voices' and the poems which went on to form *In Memoriam*, he also seems to have been drawn to poems in which he imagined what it might be like to despair and *not* recover, remaining imprisoned by delusion, melancholy, or narcotic oblivion. Croker thought that Tennyson was attracted to such subjects because he was, himself, too much a dreamer, and thus all too prone to trying to escape from reality; but the truth may well have been that these poems provided a way for him to express and explore anxieties regarding a set of psychological dangers which, at the time, he felt to be only too real.

'Porphyria's Lover' and Moral Insanity

If Tennyson was worried by doubt, then Browning was worried by certainty. As I have discussed, several of the poems Tennyson wrote during this period revolve around figures such as the speakers of 'St Simeon Stylites' and 'The Two Voices', who think (and hope) that they have experienced genuine visionary insights, but worry that they may be mistaken. Browning, by contrast, was preoccupied by figures such as Aprile and the speaker of 'Pauline': individuals who, rightly or wrongly, were completely certain of the validity of their visions and inspirations. As a result, while Tennyson's would-be visionaries question their own authority, in Browning's poems the task of questioning the authority of his speakers is passed on to the reader, instead. Indeed, one of Browning's most characteristic poetic techniques was to ventriloquise speakers who were so certain of themselves that their very certainty made them sound more than a little insane.

Nowhere is this technique shown more clearly than in 'Porphyria's Lover' and 'Johannes Agricola in Meditation', which were published in 1836 in Fox's *Monthly Repository.* In earlier mad poems of the 'Tom O'Bedlam' or 'Crazy Jane' style, the speakers usually *tell* the reader how mad they are, but in these monologues the defective mental states of the speakers must be deduced by the reader, partly from what they do and do not say and partly from how they do and do not say it. Here, for example, are the famous first lines of 'Porphyria's Lover':

> The rain set early in to-night,
>   The sullen wind was soon awake,
> It tore the elm-tops down for spite,
>   And did its worst to vex the lake:
> I listened with heart fit to break. (Browning 1983–2009, 3:250)

None of this is obviously 'mad', and the use of the pathetic fallacy, while striking, is well within the normal range of poetic expression, but the rhythms of these lines, and their patterns of sound and stress, hint that their speaker may not be of entirely sound mind. The lines are too short, for one thing; the four-stress line is standard in ballad metre, but dramatic verse usually employs the Shakespearean five-stress line, in which thoughts have sufficient space to unfold themselves. The speaker's thoughts break off too early, and this, coupled with the unusual ABABB rhyme scheme, mean that the rhyme words give the effect of coming too close together, almost tripping on one another's heels. 'Awake', 'lake', and 'break' all appear within just 26 syllables, creating an impression of a mind looping back on itself; an impression which is reinforced by alliterative runs such as 'The sullen wind was soon awake'. The overall effect is unsettling, and only becomes more so as the next few lines start breaking off in the wrong places:

When glided in Porphyria; straight
        She shut the cold out and the storm,
And kneeled and made the cheerless grate
        Blaze up, and all the cottage warm. (Browning 1983–2009, 3:250)

'And kneeled and made the cheerless grate / Blaze up' is a single thought, the kind of syntactical unit which would normally be expressed within a single, end-stopped, five-stress line, and the fact that this mundane observation is instead draped across an enjambed line break, with its terminal punctuation mark falling in the middle of a line rather than at its end, makes the speaker sound more mentally off-kilter than ever. 'Straight' and 'grate' are not logical words for the speaker to pause at, but the lineation compels a pause, however slight, after each of them, giving a halting, stumbling quality to the verse and reinforcing the impression that the speaker's thoughts will not quite fit into the places assigned for them. (Tennyson used a similar technique to communicate mental disturbance in 'The Flight', which may have been written in the same year [Tennyson 1987, 2:86–9].) These formal traits become especially pronounced in the lines which describe the speaker's murder of his lover:

That moment she was mine, mine, fair,
        Perfectly pure and good: I found
A thing to do, and all her hair
        In one long yellow string I wound
        Three times her little throat around,
And strangled her. No pain felt she;
        I am quite sure she felt no pain. (Browning 1983–2009, 3:251)

The discontinuity in the flow of the poem created by the line break in 'I found / A thing to do' highlights the discontinuity of the speaker's logic here. The 'thing to do' which he finds, namely, murder her, makes no sense in context: there is no sane reason why his discovery that Porphyria loves him should lead him to strangle her. But having 'found' it—and note the abdication of responsibility here, as though her murder was something that he chanced across passively, rather than something he actively brought about—his language suddenly becomes smooth and fluent, fitting perfectly into the spaces between the line breaks, and flowing as fluidly and evenly from 'all her hair' to 'throat around' as Porphyria's hair flows through his hands, through the triple rhyme of 'found', 'wound' and 'around' that mimics the three loops of hair around her neck. There are no further hesitations until the poem reaches the most terrible of its out-of-place line breaks—'around, / And strangled her'—in which the pause between lines becomes the momentary pause between the speaker ceasing to wind his noose of hair around Porphyria's throat and starting to pull it tight.

What I find most striking—and most horrible—about this moment in the poem is the lack of explicit build-up. The speaker is almost casual, as though Porphyria's murder was something that just happened to happen; nothing worth fussing about, especially given that—as he declares in the next line, whose staccato certainty is matched only by its improbability—'I am quite sure she felt no pain'. In this, it forms a striking contrast to the more traditional portrayal of homicidal mania in the poem upon which 'Porphyria's Lover' was loosely based: 'Marcian Colonna' (1820), by the poet (and, since 1832, metropolitan lunacy commissioner) Bryan Procter, who at the time was better known by his pen-name, Barry Cornwall (Browning 1983–2009, 3:249; Matthew and Harrison 2004, 45:454). There, once again, we find a man murdering his lover and sitting up all night with her corpse, but Marcian's crime is preceded by a much more conventional bout of insane activity:

> He clenched his hands and rushed away, away,
> And looked and laughed upon the opening day,
> And mocked the morn with shouts, and wandered wild
> For hours, as by some meteor thing beguiled.
> He wandered thro' the forests, sad and lone,
> His heart all fiery and his senses gone[.] (Cornwall 1820, 87)

In addition to all this wild laughing, shouting, rushing, and wandering, over the course of the poem the unfortunate Marcian suffers from a whole range of physical and mental symptoms of madness, including audio-visual hallucinations in which the ghosts of his ancestors tell him that he is doomed to become a murderer (Cornwall 1820, 50–52). He is described as 'a maniac, full of love, and death, and fate', and, as such, it is tragic but not exactly surprising that his story ends with his murder of his lover (Cornwall 1820, 87). In his early poem 'The Dance of Death', written in about 1826, Browning had described madness in similarly conventional terms, as characterised by 'gloomy yelling', 'a shriek as from fiends of hell', '[the] bursting eyeballs gleam', and 'the wild cry of grief and pain / Followed by the shuddering laugh': a set of symptoms which, collectively, serve to decisively mark their victim out as a stereotypical literary lunatic (Browning 1983–2009, 1:538). But it is precisely these performances of insanity which Browning so chillingly dispensed with in 'Porphyria's Lover'. Its speaker may be deluded, but he does not rant, or hallucinate, or rush about in a fit, either before or after Porphyria's death. Indeed, the poem's last lines emphasise the extent to which the speaker's actions are unaccompanied by any such traditional symptoms of insanity:

> Porphyria's love: she guessed not how
> Her darling one wish would be heard.
> And thus we sit together now,
> And all night long we have not stirred,
> And yet God has not said a word! (Browning 1983–2009, 3:252)

The speaker's beliefs appear to be deeply bizarre: he asserts that Porphyria is 'glad' that she has been murdered, and seems to think that it will have come as a pleasant surprise for her to have had 'her darling one wish' to be with her lover fulfilled by fatal strangulation. His behaviour, too, seems extremely odd: having murdered her, he sits up all night with her corpse, resting its head upon his shoulder. But instead of the lurid frenzies and hallucinations which mark Marcian out as a madman similar to the *furiosus* of eighteenth-century law, Porphyria's murderer experiences all this in calm and silence. The hallucinatory voice of God, accusing or congratulating him for his crime, is exactly what he does *not* hear.

In these respects, the speaker of 'Porphyria's Lover' provides a case study in what, at the time, was starting to be known as 'moral insanity': a form of madness whose only symptom was an extreme deviation from the moral norms of one's community (Showalter 1987, 29; Pedlar 2006, 3). This concept, which was to have great influence on the psychology and morality of the later nineteenth century, was first popularised by Prichard in his 1835 *Treatise on Insanity*, which assigned it the following definition:

> *Moral Insanity*, or madness consisting in a morbid perversion of the natural feelings, affections, inclinations, temper, habits, moral dispositions, and natural impulses, without any remarkable disorder or defect of the intellect or knowing and reasoning faculties, and particularly without any insane illusion or hallucination. (Prichard 1835, 6)

According to Prichard, 'In cases of this description the moral and active principles of the mind are strangely perverted and depraved; the power of self-government is lost or greatly impaired; and the individual is found to be incapable, not of talking or reasoning upon any subject proposed to him, for this he will often do with great shrewdness and volubility, but of conducting himself with decency and propriety in the business of life' (Prichard 1835, 4). Because moral insanity was attended with no obvious symptoms, it often went undetected: Prichard asserted that 'there are many individuals living at large, and not entirely separated from society, who are affected in a certain degree with this modification of insanity'. He lamented that it was often difficult to persuade juries to regard such people as insane, because '[if] the individual gives pertinent replies to the questions that are put to him, and displays no particular mental illusion—a feature which is commonly looked upon as essential to madness—it is most probable the suit will be rejected'. Nonetheless, Prichard insisted that such individuals really *were* mad, usually as a result of 'an hereditary tendency to madness', and that it was in everyone's best interests if they were removed into proper medical care before the 'total perversion of [their] affections' had a chance to do too much damage to those around them (Prichard 1835, 12–13).

By extending the definition of insanity to cover individuals who suffered 'no disorder or defect of the intellect', and whose only symptom was their 'strangely perverted and depraved' moral sense, the concept of moral insanity opened the way for almost all forms of immorality and criminality to be potentially understood as forms of mental pathology, even when the people thus classed as morally insane were able to give rational-sounding explanations for their actions (Prichard 1835, 6). Indeed, as Prichard remarked with obvious frustration, the morally insane individual 'has always a plausible reason to offer for his conduct', which was part of the reason why it was so hard to persuade juries to treat them like the lunatics that he believed them to be (Prichard 1835, 13). But if the category of moral insanity was accepted—as, over the course of the nineteenth century, it increasingly was—then great swathes of human behaviours previously viewed as the domain of law and religion come to be the proper territory of medical science, instead. As Suzuki has noted, the possibilities created by this change in categorisation were not lost on people at the time, and a number of families seem to have seized upon the concept of 'moral insanity' as an opportunity to have inconvenient relatives classed as insane (and thus stripped of their property rights) simply on the grounds of their odd or immoral behaviour (Suzuki 2006, 82–4).

In many respects, Porphyria's murderer seems to fit Prichard's profile of the victim of moral insanity. He is not incoherent, he suffers no mania, he experiences no 'insane illusion or hallucination', and he is capable of speaking and reasoning 'with great shrewdness and volubility', but his 'moral and active principles' are so warped that he seems to believe that murdering his lover so that they can always be together is a completely reasonable course of action (Marcian Colonna, by contrast, would be an example of '*Mania*, or raving madness') (Prichard 1835, 6). In 1836, however, this conclusion—that the speaker is mad—would probably not have been shared by everyone. As Prichard makes clear, contemporary juries were very reluctant to treat defendants as insane unless they exhibited obviously bizarre and irrational behaviour: a viewpoint reflected by the then-popular novel *Rookwood* (1834), in which a doctor declares of one villain that her behaviour was 'only excusable on the grounds of insanity, to which I wish it were possible we could attribute it', but that 'there [was] too much method in her madness to allow us to indulge any such notion' (Ainsworth 1837, 383–4). Until the institution of the M'Naghten rules in 1843, English law viewed insanity as superseding criminal guilt only if the defendant was so crazed that they had no idea what they were doing when they committed their crime, which is clearly not the case in 'Porphyria's Lover': unlike Marcian, the speaker is not in any kind of fit or frenzy at the moment of her death (Jacobs 1971, 29–32).

Whether or not Browning regarded Porphyria's lover as actually being mad when he wrote the poem in 1836, he may well have seen him as yet another example of the dangers of unrestrained imagination; and while *Paracelsus* had demonstrated how such introverted detachment from reality could be dangerous to the imaginative individuals themselves, 'Porphyria's Lover' shows that the consequences could be even more severe for those around them. There had, in fact, been a hint of this as early as *Pauline*, where in one deeply unsettling passage the speaker fantasises about murdering his beloved:

> How the blood lies upon her cheek, all spread
> As thinned by kisses; only in her lips
> It wells and pulses like a living thing,
> And her neck looks, like marble misted o'er
> With love-breath, a dear thing to kiss and love,
> Standing beneath me—looking out to me,
> As I might kill her and be loved for it. (Browning 1983–2009, 1:96)

The 'blood' which so fascinates the speaker here is the blood in Pauline's veins, made visible to him by the flush of her cheeks and lips where he has kissed her; but his choice of words, along with the proximity of his speculation that 'I might kill her and be loved for it' invites other, darker readings. His description of how 'the blood lies upon her cheek' makes it sound as though the blood is outside her body, pooling on her skin, an impression which is hardly dispelled by the speaker's descriptions of how her blood 'wells and pulses' (as it might from a wound), or how 'her neck looks, like marble' (as it might if she had grown pale from blood loss). He could almost be describing the scene of an attempted murder, in which her cheek is covered with spilt blood (probably her own) which the speaker has 'spread' and 'thinned' by kissing her blood-stained face, and in which blood 'wells and pulses' in her lips because those lips have been torn open by cuts or blows. He claims that the look he reads in her gaze, which tells him that she would love him even for murdering her, is really present in her eyes—but is it? Pauline herself, in her fictional note on the poem, gives no hint of so extravagant a passion for the speaker, and it is entirely possible that the look that he thinks he sees on her face is actually an invention of his imagination, a product of what Pauline calls his 'tournure d'esprit toute parliculière' (Browning 1983–2009, 1:89). 'Porphyria's Lover' simply takes this homicidal fantasy a step further, imagining what might result from it in the mind of someone even more 'parliculière' than the speaker of *Pauline*. Its speaker imagines he sees such a look in the eyes of his beloved, and then he does what the speaker of *Pauline* has such difficulty doing about anything: he acts on it, translating delusive imagination into grotesque reality.

'Johannes Agricola': The Dangers of Intuition

The second of what would become the 'Madhouse Cells' poems was 'Johannes Agricola in Meditation', which was marked out as a companion piece to 'Porphyria's Lover' by its formal similarities: both use four-stress lines and ABABB rhyme schemes, and each has a total length of 60 lines. As his note on the 1836 text of the poem made clear, Browning chose Johannes Agricola because of his reputation as the 'author' of an antinomian sect of the sixteenth century, and its speaker articulates a fairly standard set of antinomian doctrines: that he is one of the elect, that his election was ordained by God long before his own birth, that because of his election his salvation is assured regardless of whether he keeps or breaks the moral law, and that, because salvation is entirely independent from ordinary morality, many seemingly good and holy people will be damned while he will be saved no matter how terribly he sins along the way. It is an unusual theological position, but not an inherently irrational one; indeed, it proceeds logically from the doctrines of double predestination and justification by faith alone, both of which have historically been fairly mainstream Protestant beliefs. Nor is there anything very irrational about the speaker's language: unlike the uncomfortable sliding and stumbling of 'Porphyria's Lover', the words of 'Johannes Agricola' are smoothly and comfortably fitted to their lines. Some of his words are not too different from those one might find in a contemporary hymn:

> For I intend to get to God,
>   For 'tis to God I speed so fast,
> For in God's breast, my own abode,
>   Those shoals of dazzling glory, pass'd,
> I lay my spirit down at last. (Browning 1983–2009, 3:246)

'Johannes Agricola', then, represents a further testing of the waters regarding what actually constitutes insanity. Much of what Johannes has to say is distasteful, but having unpleasant religious beliefs is hardly the same thing as being insane. Just as the speaker in 'Porphyria's Lover' has none of the normal symptoms of homicidal madness in literature, so Johannes has none of the standard symptoms of religious mania: he doesn't make prophecies, or claim to possess special powers, or hallucinate that he is surrounded by angels and demons, or hear what he believes to be the voice of God. He articulates his views clearly, with no ranting or falling into fits; indeed, he seems to be in a much better state of mental health than Tennyson's almost-contemporary St Simeon, with his weird visions and memory lapses and blackouts. So why, in the 1842 edition, was Johannes, too, placed in the 'Madhouse Cells'?

If one accepts something like Prichard's definition of moral insanity, then it becomes much easier to class Johannes Agricola as a madman. As Sage argues, the poem enacts a collision 'between the most rigorous form of Christian self-scrutiny, and the psychopathology of what the Victorians thought of as "the criminal mind"' (Sage 1988, 96). If a thorough perversion of the moral sense is, in itself, taken to constitute a 'modification of insanity', then Johannes stands self-convicted by his own words, as when he states:

I have God's warrant, could I blend
  All hideous sins, as in a cup,
To drink the mingled venoms up,
Secure my nature will convert
  The draught to blossoming gladness fast,
While sweet dews turn to the gourd's hurt,
  And bloat, and while they bloat it, blast,
As from the first its lot was cast. (Browning 1983–2009, 3:247)

Johannes believes that, as one of the elect, even 'hideous sins' cannot prevent his progress towards salvation, while for those who are predestined to damnation even the 'sweet dews' of good deeds will only lead them to destruction. These are unappealing beliefs, but as there is no evidence that Johannes has actually *committed* 'all hideous sins', or indeed any sins at all, he can only be dismissed as a lunatic by invoking something resembling the then-novel concept of moral insanity. And yet ... something about the monologue still sounds *wrong*, somehow. Mostly this is the result of the rhyme structure, which replaces the common ABAB pattern with the ABABB rhyme scheme used in 'Porphyria's Lover': this makes every stanza seem slightly unbalanced, lurching on a little further than it really should, just as Johannes seems to have done with his theological speculations. The calm happiness with which Johannes contemplates the unearned damnation of most of the human race makes his mind a very uncomfortable place to visit, whether or not he is actually mad, and the fact that the poem employs the same verse form as 'Porphyria's Lover' implies a mental kinship between the murderer and the theologian, even if the latter has not yet done anything except meditate about his immunity to sin.

In 1836, the sanity of both speakers would have been very much up for debate—and it is precisely this debateableness which, I suspect, interested Browning in the first place. He could easily have written poems spoken by stock literary madmen of the Marcian Colonna type, whose ranting and raging would have left their lunacy in no doubt, but, instead, he penned two works whose speakers were much harder to categorise, inhabitants of what, at the time, were the contested borderlands between sanity and madness. Odd though they are, both speakers clearly consider themselves to be sane—and, as Prichard makes clear, many English juries at the time would have agreed with them. (As Suzuki has shown, the British public were extremely wary of attempts to class seemingly sane people as mad, and hostile crowds often gathered to obstruct the detention of accused lunatics who did not fit the public perception of how 'real' lunatics were supposed to behave) (Suzuki 2006, 135). This, I would argue, is part of what makes these works so much more troubling than older mad poems of the 'Mad Tom' or 'Crazy Jane' varieties. They imply that it is possible for people to slide into bizarre and dangerous mental states *without* giving way to the kind of overt performances of insanity which allow people to be straightforwardly recognised and written off as 'mad', either by themselves or by others.

Fundamentally, where both speakers have gone wrong is in neglecting Conolly's 'comparing faculty', by believing that their intuitions may be trusted without any need to measure them rationally against other sources of knowledge. The belief of Johannes that he is one of the elect is not—and, indeed, by the tenets of antinomian Christianity, *cannot* be—based on anything as fallible as human reason: he *feels* it to be so, deeply and intuitively, just as Porphyria's lover feels 'quite sure' that she welcomes her murder and that she feels no pain. These beliefs seem radically under-supported by the available external evidence, but they believe them regardless, convinced that they have access to higher sources of truth. The very fact that they communicate in monologues suggests something about their fixed certainty: being thus preoccupied with truths which are self-evident to them and invisible to everyone else, their minds have become closed circuits, immune to dialogue. In a sense, they are in 'madhouse cells' even in the 1836 texts, imprisoned within their own minds whether or not they have also been physically locked up. As a result, as Tucker has noted, these works resemble ghastly parodies of Romantic monologue-poems such as Coleridge's 'The Aeolian Harp', Wordsworth's 'Tintern Abbey', or Shelley's 'Epipsychidion', all of which also feature speakers declaring their respective spiritual insights (Tucker 1984, 124–5). Like Coleridge, Wordsworth, Shelley, and the speaker of *Pauline*, Porphyria's lover has even equipped himself with that essential Romantic prop, a silent woman to sit and listen to him while he talks. (He had to murder her first, but he seems to view that as a relatively minor detail.) These poems demonstrate the potential dangers of a life in which intuition and imagination are taken as self-evident truths, rather than being tested against any kind of external knowledge or evidence, and the risks involved in allowing one's own personal convictions to take precedence over established moral and cultural norms.

The rise of the concept of moral insanity, and the more general category of *manie sans délire* from which Prichard derived it, made it more difficult than ever for a would-be poetic visionary to know whether their intuitions could be trusted. As long as madness was understood to involve a set of positive symptoms—raving, delusions, hallucinations, manic fits, and so on—then such a person could comfort themselves with the knowledge that, as long as they exhibited none of these symptoms, they might be *wrong* but at least they were not *mad*. But if madness could strike unattended by any of these—if there was, in fact, a form of madness which consisted simply of 'perversion of the natural feelings, affections, inclinations, temper, habits, moral dispositions, and natural impulses'—then it became more important than ever for anyone whose insights and intuitions led them away from the conventional 'feelings, affections, inclinations, temper, habits, [and] moral dispositions' of the day to have other standards by which they might judge whether they were genuinely inspired rather than morally insane (Prichard 1835, 6).

Aprile, in *Paracelsus*, is a 'poor mad poet' of the traditional type, whose mad songs and hallucinations mark him out, like the madman in Shelley's 'Julian and Maddalo', as obviously mentally unsound; but Johannes Agricola is a much more ambiguous case, which serves to emphasise the importance of not allowing the faculties of imagination and intuition to act as judges of their own validity. Both Johannes Agricola and Porphyria's lover claim, like Aprile, to be motivated entirely by love: Porphyria's lover by his love of her, and Johannes by his love of God. But their love is not tempered by knowledge, or at least not the kind of objective knowledge of the outside world that Paracelsus represents: and so, when their intuition suggests that the truly loving thing to do would be to murder or to sin, it apparently does not occur to them to question it. Such reckless trust in one's own intuitions might previously have been viewed as folly, or as proof of a sinful excess of pride; but from 1835 onwards, it increasingly came to be classed as a form of madness, instead.

Unlike the deeply troubled Tennyson, whose poetry in these years was preoccupied with abnormal mental states because he himself was experiencing them so frequently, Browning had no reason to doubt his own sanity. 'Porphyria's Lover' and 'Johannes Agricola' are case studies, not confessions: indeed, their case-study nature testifies to the extent to which Browning's poetry had aligned itself with the perspective of the alienist, who listens sceptically and critically to the rhapsodies of self-proclaimed visionaries, rather than with the perspective of the visionaries themselves.

But he seems to have shared with Tennyson a growing fascination with the ways in which the subjective mind could become lost within itself, coming adrift from the outside world, and a corresponding awareness of how dangerous it might be to try to live and write guided by Shelleyan visionary imagination alone. In this he was very much in line with the medical writers of the period, who asserted explicitly what so many of Browning's poems imply through their language and poetic form: that madness could take many forms other than the old-fashioned ranting and raving which had traditionally served to characterise its literary representations. For doctors such as Prichard, mental imbalance was not always a matter of rushing about or foaming at the mouth. It could take much more subtle forms—so subtle, indeed, that even the sufferers themselves might mistake it for more trustworthy forms of religious or poetic inspiration, and not suspect their danger until it was far too late.

Embracing Uncertainty: 'Ulysses', 'Morte D'Arthur', and *In Memoriam*

By the mid-1830s, Tennyson was beginning to emerge from the depression which had engulfed him after Hallam's death. Two of his most important poems from this period, 'Ulysses' and 'Le Morte D'arthur'—the latter transparently referring to Arthur Hallam as well as the legendary king of the same name—deal directly with the necessity of continuing with life, even in the face of inevitable and irrecoverable loss. Both Ulysses and Bedivere are painfully aware that they have outlived the end of an age of heroes; their respective worlds have shrunk, 'the true old times are dead', and 'that strength which in old days / Moved earth and heaven' has departed, never to return. But in both poems they are instructed, or instruct themselves, not to abandon themselves to despair: and when Bedivere complains that 'the days darken round me, and the years', the dying Arthur tells him to live on, to 'Comfort thyself', and to pray (Tennyson 1987, 1:620, 2:17–18).

The 'true old times' which are most obviously alluded to here are Tennyson's undergraduate days at Cambridge, when the Apostles still functioned like a Round Table of Arthurian knights or a fellowship of Homeric heroes, before Robert Boyd was shot dead in Spain and Arthur Hallam succumbed to a brain haemorrhage in Vienna. But they may also refer to Tennyson's own position of historical belatedness, as a would-be

Romantic who had outlived British Romanticism proper and now found himself standing like a bewildered Bedivere in a world where not only Keats, Shelley, and Byron, but by 1834 also Hazlitt, Lamb, and Coleridge had gone, one by one, to their graves, just as 'King Arthur's table, man by man / Had fallen in Lyonesse' (Tennyson 1987, 2:5). Despite the losses that they have suffered, both Ulysses and Bedivere resolve to undertake new adventures. But unlike their journeys in 'the true old times', which had been characterised by clarity and certainty of purpose, these, tellingly, will be adventures in which they do not know where they are going.

'Ah! my Lord Arthur, whither shall I go?' cries Bedivere to his dying king, but Arthur does not answer his question; indeed, he isn't even sure where he is going himself, confessing that while he *hopes* he is being carried away to 'Avilion', 'all my mind is clouded with a doubt' (Tennyson 1987, 2:16–18). Similarly, Ulysses knows what he is sailing away from, but not what he is sailing *to*, and when he tells his crew that 'It may be we shall touch the Happy Isles', he is simply guessing at what may or may not await them beyond the edges of the map (Tennyson 1987, 1:619). These final voyages inevitably recall the earlier voyage of the Lady of Shalott, which in turn recalled those of *Alastor, Adonaïs*, and indeed Shelley himself. But when the speaker of *Adonaïs* sails 'darkly, fearfully afar', he has a clear objective in sight; he just has to follow 'The soul of Adonaïs, like a star', which 'Beacons from the abode where the Eternal are' (Shelley 1977, 406). Tennyson's Ulysses, by contrast, intends 'To sail beyond the sunset, and the baths / Of all the western stars', into waters where there will be no stars to guide him. Like Arthur and Bedivere, he simply abandons himself to hope and chance. Tennyson's poetry of 1833–1835 still depicts heroic resolution as something to be admired, but the kind of oracular insight and certainty which he had hoped for in 1830 was no longer presented as something that could be credibly aspired to.

Tennyson gradually recovered from his depression, but he did not return to the optimistic Shelleyan poetry of his undergraduate years. Then he had asserted that the true poet-prophet was an oracle indeed: that he was one whom 'angels have talked with', one who 'saw through life and death, through good and ill', and to whom 'The marvel of the everlasting will' was 'An open scroll' (Tennyson 1987, 1:243, 1:252). But in his poems of the later 1830s he credited himself with no such authority, and when he did describe apparent interactions with higher powers they were characterised by ambiguity rather than certainty. In what would become poem 69 of *In Memoriam*, Tennyson described a dream-encounter with 'an angel of the night' who speaks to him, but the result is very different from the kind of cosmic clarity which he had imagined resulting from encounters with angels in 'Armageddon' and 'Timbuctoo'. He hears the sound of the angel's voice, but 'The words were hard to understand' (Tennyson 1987, 2:384).

In poem 95, Tennyson described a trance state in which he felt his soul to be united with that of Hallam, but its culmination is ambiguous at best: as he writes, 'At length my trance / Was cancelled, stricken through with doubt', just as his King Arthur had found his mind 'clouded with a doubt' at the very moment in which he sought to affirm the reality of his own transcendence of death (Tennyson 1987, 2:413, 2:18). In poem 92, longing to communicate with Hallam's ghost, he clearly stated his dilemma:

> If any vision should reveal
> Thy likeness, I might count it vain
> As but the canker of the brain;
> Yea, though it spake and made appeal
>
> To chances where our lots were cast
> Together in the days behind,
> I might but say, I hear a wind
> Of memory murmuring the past. (Tennyson 1987, 2:409)

These were the very points which were repeatedly made, not only by contemporary writers on the supernatural such as Scott, Thacher, or Brewster, but also by writers on insanity such as Conolly: that those moments in which the visionary believed himself to '[see] through life and death' were, in the vast majority of cases, simply the results of some 'canker of the brain'. (Conolly went even further, and suggested that many such hallucinations were probably caused by nothing more than indigestion—an idea which his friend, Charles Dickens, made use of in his 1843 novel *A Christmas Carol*, in which Scrooge famously attempts to convince himself that Marley's ghost has 'more of gravy than of grave about [him]' [Tennyson 1987, 1:243; Conolly 1830, 101–5, 250; Dickens 1971, 59].) As Tennyson went on to complain, even if a ghost seemed to reveal the future, this would not suffice to prove that it was real: it could still just be his mind playing tricks on him, clothing some deduction or intuition with the hallucinated face and voice of his dead friend. Tate writes:

> These lines identify what might have been an uplifting vision of the dead as a symptom of neurological disease, and the rhyme of 'vain' and 'brain' sustains this view by hinting at the futility of trying to impose meaning on a random neuropsychological process. (Tate 2012, 107)

Ghosts appear, but they cannot be trusted. Voyages are undertaken, but their destinations are uncertain. Angels speak, but they cannot be understood. Despair recedes, but ambiguity and uncertainty replace it, and even the most rapturous trance states are 'stricken thro' with doubt'. The figure of the visionary poet-prophet, so prominent in Tennyson's earliest poetry, does not return.

Sordello: The Critique of Poetic Solipsism

Just as Tennyson's poems of the later 1830s reaffirmed his scepticism regarding the reliability of apparent instances of visionary inspiration, so Browning's *Sordello* (1840) questioned the idea that a conviction of one's own special status as an inspired visionary was a desirable precondition for being a great poet. In the first book of *Sordello*, Browning imagined himself conjuring an audience of ghosts to listen to his tale, but among this audience of imagined spirits there was one whose presence made him distinctly uncomfortable:

> [S]tay—thou, spirit, come not near
> Now—not this time desert thy cloudy place
> To scare me, thus employed, with that pure face!
> I need not fear this audience, I make free
> With them, but then this is no place for thee!
> The thunder-phrase of the Athenian, grown
> Up out of memories of Marathon,
> Would echo like his own sword's griding screech
> Braying a Persian shield,—the silver speech
> Of Sidney's self, the starry paladin,
> Turn intense as a trumpet sounding in
> The knights to tilt,—wert thou to hear! What heart
> Have I to play my puppets, bear my part
> Before these worthies? (Browning 1983–2009, 2:197–8)

The text of the poem itself gives no hint as to whom this 'pure face' belongs to, but the elucidatory headings that Browning added to the

1863 edition confirmed his identity: the 'spirit' in question was the ghost of Shelley. Just as he had in *Pauline*, Browning was emphasising the extent to which he felt his own poetic achievements fell short of the standards which Shelley had set. He declares that while he 'need not fear' the rest of the ghosts of the dead, Shelley will 'scare [him]' with his sheer superiority, robbing him of his confidence and 'heart'. For Browning in 1840, then, Shelley still defined the gold standard of true poetry. But his sense of his own relationship with his poetic predecessor had changed markedly since *Pauline*, which had ended with a plea for Shelley's spirit to remain with him forever:

> Sun-treader, I believe in God, and truth,
> And love; and as one just escaped from death
> Would bind himself in bands of friends to feel
> He lives indeed—so, I would lean on thee;
> Thou must be ever with me—most in gloom
> When such shall come—but chiefly when I die,
> For I seem dying, as one going in the dark
> To fight a giant—and live thou for ever,
> And be to all what thou hast been to me[.] (Browning 1983–2009, 1:106)

To go from imploring the spirit of Shelley to 'be ever with me' in 1832 to begging it to 'come not near' in 1840 suggests a considerable change in Browning's feelings towards his poetic predecessor. Browning's admiration for Shelley as a poet still evidently remained, but rather than viewing him as a guiding light whom he (and everyone else) should follow, he now regarded Shelley as a writer from whom he needed to maintain a certain level of distance if he was to successfully develop a poetic style of his own. In 1832 he had hoped to 'lean on' Shelley, imagining the earlier poet's poetry as supporting his own works. However, by the time he wrote *Sordello* he seems to have felt that Shelley's proximity, far from helping him to write new poetry, would only intimidate him into silence.

This shift in Browning's relationship with Shelley—who for him, as for Hallam, Mill, and Taylor, was the paradigmatic example of the more visionary and 'PHANTASTIC' side of British Romanticism—had accompanied Browning's own movement away from the kind of confessional lyric poetry he had written (and burnt) as a teenager, and towards forms of narrative and dramatic verse in which the role of the poet was not to lay bare their own heart, but to provide an insight into the minds of others.

(In his 1851 essay on Shelley, Browning would go on to describe these as 'subjective' and 'objective' poetry, respectively [Browning 1983–2009, 4:424–6].) His poetry had repeatedly explored the dangers and shortcomings of a life lived purely through the imagination, and *Sordello* proved to be no exception, presenting the would-be visionary poet not as a figure of authorial self-identification (as in *Pauline*), but as a figure observed from an ironic distance by an omniscient narrator who was thoroughly aware of his hero's many failings.

As Keenan, Yetman, and Martens have all noted, Browning's Sordello is an extremely Shelleyan figure, strongly reminiscent of the protagonist of *Alastor* (Keenan 1973, 119; Yetman 1975, 83; Martens 2011, 56–60). His secluded childhood is given over to imagination and poetic reverie, and he possesses a natural poetic talent so great that in his very first performance he is able to upstage and defeat the old poet Eglamor. Yet despite these qualifications, which should permit him to be a world-changing poet of the kind that Shelley had celebrated, Sordello in fact achieves almost nothing. His adolescent daydreams of glory go unfulfilled, and when he is finally given the chance to make a difference to the world around him, the stress of deciding what to do with it is so great that it kills him. As Yetman notes:

> [T]he only recognisably Shelleyan traits to be found in the poem are precisely those vaguely idealist habits of mind that are responsible for the hero's downfall. Beginning with the character's extreme solipsism, these would include such things as his political naïveté, as a practicing poet, his ungovernable need for public acclaim and his unrealistic expectations with regard to language; his temperamental reluctance to accommodate himself to the notion of progress and gradual change in human affairs, rather than apocalyptic leaps forward; and, in the area of personal achievement, his habitual indecision and inability to translate mental growth into meaningful action. (Yetman 1975, 83)

As with Aprile in *Paracelsus*, Sordello's devotion to a life of imagination and sensation does not equip him to change or save the world. Instead, it saps his physical and mental strength, causing him to die before he has a chance to make any meaningful impact on history. Like Aprile (and like Alastor, who is the prototype of them both), his body is left prematurely withered by his imaginative excesses, very much in line with Madden's predictions about the physically damaging side effects of literary pursuits in *The Infirmities of Genius*: after hearing Sordello speak in book 5, Taurello observes that 'poesy, sooner than politics / Makes fade young hair' (Browning 1983–2009, 2:419). Sounding very much like Shelley, Sordello insists that to be a true poet is to be 'earth's essential king', 'ordained [the world's] champion from eternity', and that by rights he, as a poet, should stand superior to a mere man of action such as Taurello; but he is forced to acknowledge that, by devoting himself to imaginative day-dreaming rather than actually working to change the world, he has squandered his poetic gifts. Even his rhapsodic, Shelleyan account of the true dignity of poets ends up falling rather flat: for when Sordello is in mid-flow, he sees with dismay that 'here the Chief [i.e. Taurello] immeasurably yawned' (Browning 1983–2009, 2:425–9).

As its preliminary praise of Shelley suggests, *Sordello* does not entirely reject the figure of the poet-visionary. Indeed, in book 1, the narrator praises those poets who are able to attain genuine visionary perception, and in whose eyes the world becomes charged with the grandeur of the divine:

> Up and down
> Runs arrowy fire, while earthly forms combine
> To throb the secret forth; a touch divine—
> And the scaled eyeball owns the mystic rod;
> Visibly through his garden walketh God. (Browning 1983–2009, 2:219)

Sordello, however, fails to become this kind of poet-mystic. His 'eyeball' is not 'scaled'—which here means 'stripped of its scales', as Saul's were on the road to Damascus—and his failure is due to the fact that he does not love the world's beauty for itself, but only as a kind of reflection of what he takes to be the glory of his own mind:

> For there's a class that eagerly looks, too,
> On beauty, but, unlike the gentler crew,
> Proclaims each new revealment born a twin
> With a distinctest consciousness within,
> Referring still the quality, now first
> Revealed, to their own soul—its instinct nursed
> In silence, now remembered better, shown
> More thoroughly, but not the less their own. (Browning 1983–2009, 2:221)

Because the members of this 'class' believe that each beauty they perceive in the world is merely the 'twin' of something that already exists within themselves, they feel no need to actually put their powers to use. They feel their souls to be already 'rife / With grandeurs' and 'equal to being all'; so what they do or do not actually achieve during their physical life on earth is of strictly secondary importance, a matter of 'trivial accidents', and their abilities thus often remain 'Dormant'. Due to their self-satisfaction with their own supposed genius, the beliefs of such poets rather come to resemble an artistic version of the antinomianism of Browning's Johannes Agricola. Just as he believed that the certainty of his election made it irrelevant whether or not he ever actually performed any good deeds, so they believe that the certainty of their spiritual greatness makes it irrelevant whether or not they ever actually write any poetry.

Browning's diagnosis of this 'class' of poets seems quite amazingly specific—how many would-be poets can ever have had this exact combination of beliefs and attributes?—which points, again, to how deeply embroiled *Sordello* is with the specific temptations and dangers attendant upon what Browning saw as the Shelleyan approach to poetry. In writing these lines, Browning may have had in mind stanzas 44–46 of *Adonaïs*, in which Shelley described the ascent of the spirit of Keats to join 'The inheritors of unfulfill'd renown'. There, Shelley asserts that 'When lofty thought / Lifts a young heart above its mortal lair', death cannot blot out the resulting brightness; and even if the 'young heart' in question ceases to beat without acquiring any kind of earthly fame, its power continues to live on in 'the Unapparent'.

> And many more, whose names on Earth are dark,
> But whose transmitted effluence cannot die
> So long as fire outlives the parent spark,
> Rose, robed in dazzling immortality. (Shelley 1977, 403)

Elsewhere in *Adonaïs*, Shelley stresses that it is because of his real artistic achievements that Keats will be immortal: he has become 'a portion of the loveliness / Which once he made more lovely'. But if the crucial qualification is understood to be 'lofty thought' rather than actual poetic practice, then these stanzas leave open the possibility that immortality would have been his lot even if he had never written a line; he might have transmitted a beneficial 'effluence' into the world simply by *being himself*. In this, *Adonaïs* is very much in keeping with the earlier *Alastor*, whose protagonist is repeatedly referred to as 'the Poet' even though he never actually seems to write any poetry: possessing a 'high and holy soul', and wandering the world singing 'passionate notes' and thinking 'high thoughts' are apparently sufficient to qualify him for the title (Shelley 1977, 72–3, 85, 402). The 'class' of poets to which Sordello belongs, with their indifference to actual poetic attainment, would thus be those who were led by such beliefs to conclude that what they did or did not achieve during life was irrelevant, because the quality of their poetic souls alone was enough to guarantee their transcendence. It is not a class which one would expect to have been numerous in thirteenth-century Italy, or indeed virtually anywhere except nineteenth-century Britain.

In these lines, then, Browning can be seen weighing up the risks and opportunities of such an approach to poetry. At its best, it can lead to genuine visionary perception, in which the world is charged with 'a touch divine'; but at its worst it can lead away into solipsistic self-regard, in which thinking grand poetic thoughts—or what Sordello believes to be grand poetic thoughts, at any rate—is allowed to substitute for the actual practice of art (Browning 1983–2009, 2:219). Like Taylor before him, Browning ultimately emphasises the need for true engagement with the external world: not merely as a 'revealment' of the poet's real or imagined inner qualities, but as something genuinely outside or beyond the self. Sordello belongs not to 'the gentler crew' of true visionaries, but to this 'class that eagerly looks'; probably the speaker of *Pauline* did, too, as the claims of this 'class' to possess 'a distinctest consciousness within' are strongly reminiscent of his assertion to be 'made up of an intensest life, / Of a most clear idea of consciousness' (Browning 1983–2009, 1:42). As a result, Sordello's poetic project is doomed from the start, and his artistic powers are fated to remain 'Dormant' rather than being deployed to change the world in which he lives.

Conclusion: Abandoning the Oracles

In his essay on 'The Present State of Poetry', published in 1838, Bulwer-Lytton singled Tennyson out as an example of the extent to which 'the eyes of our rising generation are yet too much dazzled by the lustre of their immediate predecessors'. British poetry, he argued, had never really recovered from the shock of Byron's death, and 'Young men of genius', such as Tennyson, were still imaginatively enthralled by the previous generation, with the result that 'their inspiration can be too distinctly traced to oracles with whose devices we have grown familiar'. Byron, of course, was one of these 'oracles', but as Bulwer-Lytton wrote, 'more apparent, haunting, and oppressive, appear the influence of Wordsworth and of Shelley'. Yet he was far from convinced that these were healthy models: he compared the influence of Byron to that of an 'opiate', and thought Shelley so very abnormal that those who imitated him would inevitably 'contract affectation' (Bulwer-Lytton 1841, 1:337–41).

Tennyson, in Bulwer-Lytton's view, was in danger of being badly misled by this 'haunting', which he implicitly associated with mental and physical ill-health. Scott 'went the right way to the human heart', whereas Shelley 'lacked masculine and robust simplicity'. By setting themselves up as 'oracles', raised above—and perhaps even actively opposed to—the ebb and flow of everyday experience, such poets risked leading their imitators astray into abnormal and unhealthy poetic practices. As he wrote disapprovingly of Shelley, in words which recall Felluga's observation that Romantic poetry was often associated with sexual perversity during the nineteenth century, 'what was most natural to that strange thinker and most fanciful idealist is often what to ordinary persons may seem the least natural' (Felluga 2005, 140; Bulwer-Lytton 1841, 1:339–44). Tennyson, he concluded, would be well-advised to seek his models elsewhere.

If Bulwer-Lytton had been able to read the poems which Tennyson was actually writing at the time, however, he might have been rather less concerned about him: for by 1838, Tennyson had left the Shelleyan 'haunting' of his undergraduate years far behind. He no longer sought to model himself upon such 'oracles', still less to set himself up as an oracle in his own right. Instead, just as contemporary writers on madness had recommended, his poems from 'St Simeon Stylites' to *In Memoriam* depicted such oracular insights and pronouncements with great scepticism, and demonstrated a healthy respect for the deceptive powers possessed by 'canker[s] of the brain' (Tennyson 1987, 2:409).

A comparable progression away from sub-Shelleyan Romanticism can be seen in the works of Browning over the same years. Taken as a whole, the poetry written by Browning between 1832 and 1840 suggests that his initial devotion to an ideal of the poet as inspired visionary gave way to a deep distrust of the poetry of pure subjectivity as the decade wore on. In poem after poem, the reliance on imagination and intuition at the expense of knowledge, judgement, and engagement with external reality is critiqued as physically draining, politically ineffectual, and potentially dangerous to one's sanity. Horne, Mill, and Hallam, building on Shelley, had developed the idea that being a great poet was pre-eminently a matter of having a certain kind of mind; only possess a sufficiently 'poetic *nature*', and all the rest would follow (Mill 1981, 1:355). But in *Sordello*, as in *Paracelsus*, this turns out to be quite untrue. The reality behind the legend of the poet-hero Sordello is revealed to be a man whose trust in his poetic genius, and consequent reliance upon pure imagination, leaves him so crippled in body and mind that he fails politically, poetically, and ultimately even physically, his weakened constitution unable to endure the strain of making a single decision of genuine importance.

What underlies all these works, from *Pauline* to *Sordello*, is an increasing wariness of taking states of apparent inspiration at face value. Johannes Agricola's knowledge that God will send him to heaven even if he commits every sin comes from apparent inspiration; so does the belief of Porphyria's lover that the best way for him to reward Porphyria's love is by murdering her, and Sordello's conviction that his soul is so great that simply daydreaming about heroism is enough to earn him immortality. They are all, like Aprile, more or less lost in solipsistic mazes, which gradually lead them into states that either resemble or constitute actual insanity. In this, the poems in which they appear are very much of a piece with the overall tenor of the medical psychology of the 1830s, with its deepening suspicion of all unusual states of consciousness. For Browning, as for Conolly, it was only by constantly testing imagination and inspiration against real knowledge of the external world, avoiding all forms of excessive inwardness, that such pitfalls could be avoided and the sanity of the would-be poet or prophet maintained. The last years of the 1830s provided some spectacular examples of just how dangerous the consequences of failing to maintain this mental balancing-act could be, and it is these cases which form the subject of my next chapter.

References

Primary Sources

Ainsworth, William Harrison. 1837. *Rookwood: A Romance*. London: Richard Bentley.

Browning, Robert. 1983–2009. *Poetical Works*, ed. Ian Jack, Rowena Fowler, Margaret Smith, Robert Inglesfield, et al., 15 vols. Oxford: Clarendon.

Bulwer-Lytton, Edward. 1841. *Critical and Miscellaneous Writings*. 2 vols. Philadelphia: Lea and Blanchard.

Coleridge, S.T. 1999. *Poems*, ed. John Beer. London: David Campbell.
Conolly, John. 1830. *An Inquiry Concerning the Indications of Insanity*. London: John Taylor.
Cornwall, Barry [Bryan Procter]. 1820. *Marcian Colonna, an Italian Tale, With Three Dramatic Scenes, and Other Poems*. London: John Warren and C & J Ollier.
Dickens, Charles. 1971. In *The Christmas Books*, ed. Michael Slater, vol. 1. Harmondsworth: Penguin.
Mill, John Stuart. 1981. In *Collected Works*, ed. John Robson. Toronto: University of Toronto Press.
Milton, John. 1998. In *The Riverside Milton*, ed. Roy Flannagan. Boston: Houghton Mifflin.
Morison, Alexander. 1828. *Cases of Mental Disease*. Edinburgh: Maclachlan and Stewart.
Prichard, James Cowles. 1835. *A Treatise on Insanity and Other Disorders Affecting the Mind*. London: Sherwood, Gilbert, and Piper.
Shelley, Percy Bysshe. 1977. In *Poetry and Prose*, ed. Donald Reiman and Sharon Powers. New York: Norton.
Taylor, Henry. 1834. *Philip Van Artevelde*. 2nd ed., 2 vols. London: Edward Moxon.
Tennyson, Alfred. 1982. *Letters*, ed. Cecil Lang and Edgar Shannon, Jr., 3 vols. Oxford: Clarendon.
———. 1987. *The Poems of Tennyson*, ed. Christopher Ricks, 2nd ed., 3 vols. Harlow: Longman.

Secondary Sources

Baker, John. 2004. *Browning and Wordsworth*. Madison: Fairleigh Dickinson University Press.
Batchelor, John. 2014. *Tennyson: To Strive, To Seek, To Find*. London: Vintage.
Colley, Ann. 1983. *Tennyson and Madness*. Athens, GA: University of Georgia Press.
Colville, Derek. 1970. *Victorian Poetry and the Romantic Religion*. Albany: State University of New York Press.
Cronin, Richard. 2012. *Reading Victorian Poetry*. Oxford: Wiley-Blackwell.
Faas, Ekbert. 1988. *Retreat into the Mind: Victorian Poetry and the Rise of Psychiatry*. Princeton: Princeton University Press.
Felluga, Dino. 2005. *The Perversity of Poetry: Romantic Ideology and the Popular Male Poet of Genius*. New York: New York State University Press.
Jacobs, Francis. 1971. *Criminal Responsibility*. London: Weidenfield and Nicolson.
Jump, John, ed. 1967. *Tennyson: The Critical Heritage*. London: Routledge and Kegan Paul.

Keenan, Richard. 1973. Browning and Shelley. *Browning Institute Studies* 1: 119–145.

Kennedy, Richard, and Donald Hair. 2007. *The Dramatic Imagination of Robert Browning: A Literary Life*. Columbia: University of Missouri Press.

Lourie, Margaret. 1979. Below the Thunders of the Upper Deep: Tennyson as Romantic Revisionist. *Studies in Romanticism* 18 (1): 3–27.

Martens, Britta. 2011. *Browning, Victorian Poetics, and the Romantic Legacy: Challenging the Personal Voice*. Farnham: Ashgate.

Matthew, H.C.G., and Brian Harrison, eds. 2004. *Oxford Dictionary of National Biography*. 60 vols. Oxford: Oxford University Press.

Pedlar, Valerie. 2006. *The Most Dreadful Visitation: Male Madness in Victorian Fiction*. Liverpool: Liverpool University Press.

Platizky, Roger. 1989. *A Blueprint of His Dissent: Madness and Method in Tennyson's Poetry*. Lewisburg: Bucknell University Press.

———. 2002. "Like Dull Narcotics, Numbing Pain": Speculations on Tennyson and Opium. *Victorian Poetry* 40 (2): 209–215.

Richardson, Alan. 2001. *British Romanticism and the Science of the Mind*. Cambridge: Cambridge University Press.

Ricks, Christopher. 1989. *Tennyson*. 2nd ed. Berkeley: University of California Press.

Rivers, Charles. 1976. *Robert Browning's Theory of the Poet, 1833–1841*. Salzburg: Universität Salzburg.

Sage, Victor. 1988. *Horror Fiction in the Protestant Tradition*. New York: St Martin's Press.

Shannon, Edgar. 1981. Poetry as Vision: Sight and Insight in "The Lady of Shalott". *Victorian Poetry* 19 (3): 207–223.

Showalter, Elaine. 1987. *The Female Malady: Women, Madness, and English Culture, 1830–1980*. London: Virago.

Stevenson, Catherine Barnes. 1982. The Shade of Homer Exorcises the Ghost of De Quincey: Tennyson's "The Lotos-Eaters". *Browning Institute Studies* 10: 117–141.

Suzuki, Akhito. 2006. *Madness at Home: The Psychiatrist, the Patient, and the Family in England*, 1820–1860. Berkeley: University of California Press.

Tate, Gregory. 2012. *The Poet's Mind: The Psychology of Victorian Poetry 1830–1870*. Oxford: Oxford University Press.

Tennyson, Charles, and Hope Dyson. 1974. *The Tennysons: Background to Genius*. London: Macmillan.

Tucker, Herbert. 1984. From Monomania to Monologue: "St. Simeon Stylites" and the Rise of the Victorian Dramatic Monologue. *Victorian Poetry* 22 (2): 121–137.

———. 1988. *Tennyson and the Doom of Romanticism*. Cambridge, MA: Harvard University Press.

Whitehead, James. 2017. *Madness and the Romantic Poet: A Critical History*. Oxford: Oxford University Press.

Wood, Jane. 2001. *Passion and Pathology in Victorian Fiction*. Oxford: Oxford University Press.

Yetman, Michael. 1975. Exorcising Shelley Out of Browning: "Sordello" and the Problem of Poetic Identity. *Victorian Poetry* 13 (2): 79–98.

CHAPTER 4

Sir William's Last Stand: Poetry and Insanity in England, 1837–1842

This chapter deals with the years 1837 to 1842, which saw the rise of the British asylum system, and of the new medical psychology which supported it. For most of this period, both Alfred Tennyson and John Clare were residents of the village of High Beach—the former as a friend and business associate of the local asylum-keeper, Matthew Allen, and the latter as one of his patients—and their writings on poetry and madness are discussed in light of their respective relationships with the burgeoning asylum system. Finally, it considers the bizarre case of John Tom, who was released from Kent County Lunatic Asylum in 1837 and went on to lead a doomed uprising against the government in 1838, claiming to be Jesus Christ reborn. Tom's rebellion, and its aftermath, will be used to explore the ever-wider ways in which insanity was used to explain unusual and disruptive behaviour during the later 1830s, and the consequences that this had for poets such as Tennyson and Clare.

High and Dry: Clare and Tennyson at High Beach

The years 1837–1840, during which both Alfred Tennyson and John Clare resided in the tiny village of High Beach just outside London, feel like one of the great literary near misses of the nineteenth century. Clare was first confined in Dr. Allen's private madhouse at High Beach in 1837, after suffering for some years from depression and paranoid delusions, probably aggravated by heavy drinking and financial anxieties (Bate 2004,

© The Author(s) 2019

J. Crawford, *Inspiration and Insanity in British Poetry*,
Palgrave Studies in Literature, Science and Medicine,
https://doi.org/10.1007/978-3-030-21671-9_4

407–8). Later that year, Tennyson arrived in High Beach in search of a new home, his family having been required to leave the Somersby rectory after his father's death. Clare was permitted some freedom to roam about outside the asylum, and in a village as small as High Beach it would be surprising if the two poets had not crossed paths from time to time. Later, in 1840, Tennyson became the friend and business partner of Dr. Allen, who took him on at least one visit to his asylum. Tennyson was apparently 'delighted with the mad people'; was John Clare not amongst them? (Tennyson 1982, 1:183n). Did Allen never mention to his new friend that one of his patients was a poet, a man whose works had been much praised back in the 1820s, and who was even now continuing to write great numbers of verses? Did neither man ever notice the other as a fellow East Midlander? (Somersby, where Tennyson grew up, was only 30 miles from Clare's home village of Helpston.) Could Clare and Tennyson really never have met?

'It is not known whether Alfred Tennyson met Clare while he was in Matthew Allen's care, but it seems possible that he would have done', writes Tennyson's biographer, John Batchelor. 'They are almost certain to have encountered each other', writes Clare's biographer, Jonathan Bate (Batchelor 2014, 108; Bate 2004, 432). Perhaps they did—but, if so, there seems to be no record of it. No mention of Clare from Tennyson, no mention of Tennyson from Clare: one of the first of the great Victorians and one of the last of the great Romantics seem to have missed one another by inches. Yet their shared residence at High Beach, and their shared acquaintance with Dr. Allen, remains curiously suggestive. From the outside, the two men must have seemed to their High Beach neighbours to have little in common: Tennyson was a young gentleman of leisure, while Clare was an ageing, impoverished labouring-class madman. Tennyson was able to enter into business with Dr. Allen as an equal; Clare was just one of Allen's patients. But, in fact, during his time at High Beach, Tennyson was painfully aware of just how tenuous both his sanity and prosperity might be, and his interactions with Allen would end up leaving his grasp on both shakier than ever. He would ultimately become rich, famous, and successful: but in 1837 he was not yet any of those things, and it must have seemed entirely possible to him that he might yet end up like John Clare, instead.

Indeed, if Tennyson *had* met Clare, he would probably have found many aspects of his case worryingly familiar. Clare suffered from depression and paranoid delusions, worsened by his heavy drinking; in his poem

'The Sequel to John Barleycorn', written while he was in Allen's asylum, he reflected on the damage which drink had done to him, ruefully concluding that 'He [i.e. drink] ground me on, he kept me down / And made me no amends' (Clare 1984, 1:24; Bate 2004, 409–15). Tennyson's alcoholic father never seems to have experienced the sort of elaborate delusions which increasingly plagued Clare, although he had been something of a fantasist: but his combination of drink and depression had blighted the childhoods of Tennyson and his siblings, and by 1837, Tennyson's brothers were showing an alarming tendency to follow in their father's footsteps. Edward Tennyson was still confined to his asylum and showing no signs of improvement, and Septimus Tennyson's depression had grown so severe that he voluntarily submitted himself to Allen's care several times during the family's years at High Beach (Ricks 1989, 21–4; Batchelor 2014, 107–8). Charles Tennyson had started taking such large doses of laudanum that by the mid-1830s his relatives were speculating that his addiction might prove fatal, while his younger brother Arthur was drinking so heavily that in 1843 he, too, voluntarily entered a private asylum in a last-ditch attempt to break off what his mother described as his 'ruinous and destructive habit of drinking' (Tennyson 1982, 1:87n, 1:107n, 1:217; Stevenson 1982, 134–6). Alfred did not drink so heavily as Arthur, did not take so much opium as Charles, and did not feel as crazy as Edward or as depressed as Septimus. But he *did* drink, take opiates, and suffer from melancholy, and sometimes he feared for his own sanity. By the time of his visit to Allen's asylum in 1840, he was in a very low mood—and the total failure of Dr. Allen's business schemes, which cost Tennyson most of the money that he had inherited from his grandfather, was to push him lower still. From 1840 to 1845 he was depressed, miserable, a prey to hypochondria, and far too melancholy to write (Ricks 1989, 163–8).

As I have discussed, in the 1830s poetry was understood to be a psychologically dangerous career to pursue. In 1837, the same year that Clare and Tennyson moved to High Beach, one of the architects of Britain's new asylum system, W.A.F. Browne, argued that priests and poets were especially vulnerable to insanity because they 'arouse all our most energetic emotions' (Scull 1991, 58). 'Vex not thou the poet's mind; / For thou canst not fathom it,' Tennyson had written, grandly, at the height of his Shelleyan enthusiasm back in 1830: but Clare in 1837 presented melancholy proof that when the minds of poets became truly unfathomable to other people, they did *not* cease to 'vex' them in the least. No one could 'fathom' Clare's insistence that he was actually Lord Byron, or his belief that he was simultaneously married to his real wife, Patty, and his childhood sweetheart, Mary, the latter of whom had in fact been dead for years. But rather than respecting his 'poet's mind' as 'holy ground', as Tennyson had commanded in 'The Poet's Mind', Clare's doctors proceeded to 'vex' him considerably. Back in 1830, Tennyson had described what would happen if a 'dark-brow'd sophist', out of sympathy with the unfathomable mind of the poet, were to trespass there:

> The flowers would faint at your cruel cheer.
> In your eye there is death,
> There is frost in your breath
> Which would blight the plants.
> Where you stand you cannot hear
>   From the groves within
>   The wild-bird's din.
> In the heart of the garden the merry bird chants,
> It would fall to the ground if you came in. (Tennyson 1987, 1:247)

The verse is clumsy, but the sentiment is earnest, demonstrating that even at the very outset of his poetic career Tennyson understood that the capabilities of 'The Poet's Mind' could be desperately vulnerable to outside meddling. The visionary insights of the poet are not, it seems, enough to preserve them from the interference of an uncomprehending and unsympathetic audience, in whose presence 'the merry bird' of the poetic voice falls mute and drops down dead.

A few years later, Clare had described his own worsening state of depression and mental alienation in very similar terms of wilting vegetation and killing frost, writing of how 'I found the pleasure past and a winter come at last' (Clare 2004, 260). For Tennyson in 1830 these dangers were still largely hypothetical, but by 1837 they had become all too real for Clare. He saw himself as a victim, not of an individual 'dark-brow'd sophist' like the one imagined by Tennyson, but of the whole social system which such sophists had designed: a system which respected nothing except wealth, and consequently scorned and neglected him because of his poverty. Tennyson's situation was much less precarious than Clare's, but by 1840 he had experienced his own share of difficulties due to his lack of wealth: Hallam had been prevented from marrying his sister, Emily Tennyson, by their lack of money, and in 1840 Tennyson himself had been compelled to break off his engagement to Emily Sellwood for the same reason. Years later, his brother Charles recalled that he did this because 'his monetary position did not improve as he expected, and he got into low and desponding spirits as to his future prospects' (Batchelor 2014, 110). His preoccupation with the figure of Mariana, the subject of two of his early poems, may have reflected his awareness of the ways in which his life might be limited by his unstable financial situation: in *Measure for Measure*, it is Mariana's lack of a dowry which leads Angelo to refuse to marry her, abandoning her to a life of humiliated solitude in her 'moated grange'. Like the farmers, trees, and animals that Clare lamented in his poems, it was to 'self interest' that Mariana 'fell a prey', leading her to decline, like Clare himself, into depression and despair—a theme which Tennyson would later explore more fully in 'Maud' (Clare 2004, 260).

Such disappointments could potentially have extremely severe psychic consequences, as Tennyson's new neighbour, Dr. Allen, knew only too well. In his *Essay on the Classification of the Insane* (1837), Allen had described a real-life Mariana whom he encountered at the York asylum:

> She was from home when her mind received a severe shock by the unexpected intelligence of her father having put an end to his own existence. Soon after this a grievous disappointment completed the overthrow of her mind. Before her father's death, it was generally supposed he was wealthy: she was then engaged to one who had secured her affections; after her lover knew of her father's death, and the involved state of his affairs, he still continued to profess his attachment, and held out the prospect of speedily fulfilling his promise of marriage;—she believed him, until she happened accidentally in company to cast her eye on the announcement of his marriage to another, when she shuddered and shrieked, and exclaimed "Wretch!" and from that moment she was insane, and has been so ever since. (Allen 1837, 188–9)

Sudden financial reverses featured in a number of Allen's other anecdotes and case studies of the insane. Elsewhere in the *Essay*, he noted:

> The greatest number of those who become insane, become so between the ages of thirty and forty,—a period when establishments are formed, and habits have been strengthened by time, while the feelings yet retain all their energy and susceptibility of action. Thus we can conceive why reverses and disappointments should then have the most fatal and overwhelming influence. (Allen 1837, 147)

In January 1840, when his engagement with Emily Sellwood was broken off, Tennyson had arrived at the dangerous age of 30 just five months before; and as he became increasingly involved in Allen's business plans, it must have been clear to the doctor that some care would be needed if the 'disappointments' he had suffered were not to 'have the most fatal and overwhelming influence' on the poet's mind.

'The Nature of the Invention': The Schemes of Dr. Allen, 1840–1842

Both Tennyson and Clare could fairly say that money lay at the root of their respective histories of melancholia. Clare had suffered from depression since he was a young man, but in the years leading up to his confinement much of his gloom was due to his financial hardships: he had failed as a farmer, and he was unable to make money by selling his poetry. Allen himself thought that all Clare really needed to set his mind at ease was a steady income, declaring in a letter which he wrote to *The Times*:

> I had then not the slightest hesitation in saying that if a small pension could be obtained for him, he would have recovered instantly, and most probably remained well for life. I did all I could to obtain it for him, but without the slightest success. (Bate 2004, 430)

Money had been the source of the Tennyson family's problems, too. His father, Charles, had taken to drink after *his* father had decided to make his younger son his heir, thus forcing Charles into an unwanted church career. That his subsequent alcoholism and depression had taken a heavy toll upon his children was, by 1837, all too obvious; and while the inheritance that Tennyson received in 1835 had allowed him to avoid following his father into the church, by 1840 it was not at all clear that it would prove sufficient for him to continue living as a literary gentleman of leisure. Very plainly it had not been enough for him to be able to marry Emily Sellwood.

It was at this stage that Dr. Allen intervened in the lives of both poets. Described by Cooter as 'a spendthrift, inveterate liar, and persuasive personality', he had enjoyed more success in his chosen career than either Tennyson or Clare; having entered the 'trade in lunacy' as an apothecary at the York Asylum in 1819, he had gone on to become a successful lecturer in phrenology, before rising to become the proprietor of his own up-market private madhouse (Cooter 1984, 153, 273; Batchelor 2014, 107n). The late 1830s, however, were years in which the 'trade in lunacy' was changing fast; and it is clear from Allen's 1837 *Essay* that he feared that the recent legislation and parliamentary debates on lunacy, which included a good deal of criticism of private madhouses, might have a negative impact upon businesses such as his own (Allen 1837, 60–1, 106–7). It may thus have been concern over the long-term viability of the madhouse trade which led him, in 1840, to seek other forms of income.

Allen's solution was to found a company, provisionally dubbed 'Pyroglyph' which would carry out decorative wood-carving work by means of steam-powered machinery. All he needed, he believed, was sufficient capital; and he thus persuaded Tennyson to invest his entire inheritance in the business, along with a great deal of additional money from other members of his family, assuring him that the profits realised would be sufficient to make them all financially independent for life (Batchelor 2014, 129–31). As if offering such assistance to one poet wasn't enough, Allen also started to write to potential benefactors, trying to raise money for a subscription edition of Clare's poetry in order to pay for his maintenance at High Beach—although as the money raised went to Allen rather than Clare, it, too, probably ended up being invested in the wood-carving business (Bate 2004, 429–32).

The Tennysons entered into his partnership with Allen in high hopes of success. Alfred wrote in March 1841 that 'the concern I believe is going on very well', while in October his brother Charles explained the situation to their sceptical uncle:

> This same speculation would not have been thought of being entered into, had not there been every reason both from the nature of the Invention and the character of the English Patentee [i.e. Allen] to warrant them [i.e. his sisters] in their intention of taking a share. (Tennyson 1982, 1:190, 1:196)

Clare, mad though he might have been, was less naive: in 'All[e]ns hells', he wrote, 'They'll cheat you of your money friend / By takeing [sic] too much care o't' (Clare 1984, 1:36). Allen's agent proved untrustworthy, and Allen's letters to Tennyson about the progress of the 'Pyroglyph' scheme seem, at best, to have been liberal with the truth (Tennyson 1982, 1:197; Batchelor 2014, 129). The plan was an utter failure, and all the money invested was lost.

The stress and anxiety caused by his disastrous dealings with Allen aggravated Tennyson's depression. His health gave way, and he spent the early 1840s in and out of hydropathic establishments, writing almost nothing. As Colley notes, 'the line between himself and the inmates of Allen's institution was growing thinner and thinner' (Colley 1983, 53–4). In July 1842, having learned that the scheme in which he had invested so much was likely to fail, he wrote to Allen: 'My spirits have been dejected, my nerves shattered, my health affected by what I have heard.' That September he wrote to Edmund Lushington: 'What with ruin in the distance and hypochondriacs in the foreground I feel very crazy. God help all.' Nor was this merely a passing mood: a year and a half later he wrote to Fitzgerald that he had been 'left a beggar', and that 'the perpetual panic and horror of the last two years has steeped my nerves in poison' (Tennyson 1982, 1:205, 1:210, 1:222). It was one of the lowest points in Tennyson's career, and perhaps the moment at which he came closest to sharing Clare's unhappy fate.

In 1840, then, Allen was raising money *from* Tennyson, and also raising money *on behalf of* Clare. The failure of their paths to cross thus starts to seem even odder: did Allen never ask his new business partner whether, as well as the 2000 pounds he was investing in the Pyroglyph scheme, he would be able to spare a guinea or two for the support of a suffering fellow poet? Perhaps they simply happened never to meet; or perhaps they met, but the meeting made no impression upon either of them. But I wish to suggest a third possibility: perhaps Allen deliberately kept them apart.

This may seem somewhat fantastical; but the early nineteenth-century madhouse was a fantastical place, in which all kinds of deceptions could be carried out in the name of treating the insane. When Edward Tennyson was sent to the York Asylum, his uncle planned to tell him that he was going there to study medicine: it was only once he arrived that he was to learn that he was not going to be a doctor, but a patient (Tennyson 1982, 1:83). In his *Essay*, Allen stated that it was important for mad-doctors not to lie to their patients, but he seems to have had a very flexible idea of what constituted telling them the truth: he goes on to boast that on one occasion he kept a suicidal patient under continual close observation at High Beach for six months without once allowing him to realise that he had, in fact, been confined to a private madhouse, instead keeping up the pretence that he and his staff were merely friends of the patient's parents who had invited him to stay for a while (Allen 1837, 29, 44, 96). (In 1871, Le Fanu used such a scenario as the foundation of his Gothic novel *The Rose and the Key*, whose heroine does not find such a course of systematic deception to be at all beneficial to her mental state.) Furthermore, Allen has a good deal to say about the importance of keeping madhouse patients away from people who are likely to influence them for the worse: thus depressive 'torpid cases' like Clare (and Tennyson?) were best cured by placing them amongst those 'in a more lively state', rather than allowing them to drag one another down in mutual misery (Allen 1837, 49).

When asylums were visited by outsiders, as High Beach was by Tennyson in 1840, Allen insisted that 'nothing can be worse' than allowing them to talk to the inmates about 'the subject[s] of their individual insanity': such conversations, he maintained, can seriously aggravate the condition of such unfortunates, who should instead be artfully steered by their doctors towards topics which would make them realise, without the need for any kind of confrontation, that their delusions must be false (Allen 1837, 45–6, 100–1). The mad, he repeatedly stated, were like children, and had to be manipulated for their own benefit (Allen 1837, 44–6, 68, 103). Confronted with the case of John Clare, driven into depression by poverty and poetry, and suffering from the delusion that he was actually Lord Byron (and possibly Shakespeare as well), it thus seems possible to me that Allen would have felt it best to limit his contact with three sorts of people: depressives, people likely to talk to him about financial problems, and people likely to talk to him about poetry. Tennyson fell into all three categories, and Allen may well have feared that introducing him to Clare as a fellow poet would have been tantamount to an invitation for Clare to start explaining that he was really Byron, that he was working on a new canto of *Don Juan*, and so on, thus potentially retarding or endangering his recovery by focusing his mind on those subjects concerning which he was most deeply deluded. Seeing Tennyson's own rather fragile mental state, Allen may also have felt that keeping the two poets might have been best for Tennyson, as well, for fear that the spectacle of the older poet's madness and melancholy might adversely affect a man he knew to be already struggling with 'reverses and disappointments' of his own. In addition, when Tennyson visited the asylum to see the patients in 1840, Allen badly needed Tennyson's money for his 'Pyroglyph' scheme. The last thing that he would have wanted would have been for some unfortunate encounter to make his potential business partner collapse into the kind of helpless depression for which Allen had been treating his brother Septimus.

If he *had* decided to keep Tennyson and Clare apart, Allen would probably have done so secretly, in accordance with his stated beliefs about the best ways of managing the insane. This could account for the lack of any mention of Clare by Tennyson: rather than actually explaining that he wanted to keep them from meeting, Allen may simply have neglected to mention his presence at High Beach, and steered him out of the way when Tennyson came to visit the asylum. If this *is* what happened, then even if Allen's motives were essentially benevolent, his actions may well have done a further disservice to both men. With their poetic talents and their shared love of nature poetry, Clare and Tennyson had much more in common than just histories of disappointment and depression: indeed, they even seem to have noticed the same flowers in their walks though Epping Forest. In a poignant poem of 1841, interpolated into the angry satire of his 'Don Juan a Poem', Clare wrote:

> Eliza now the summer tells
> Of spots where love & beauty dwells
> Come & spend a day with me
> Underneath the forest tree
> Where the restless water flushes
> Over mosses mounds & rushes
> & where love & freedom dwells
> With orchis flowers & fox glove bells
> Come dear Eliza set me free
> & oer the forest roam with me. (Clare 1984, 1:95–6)

'Eliza' did not come to set Clare free, and may not have existed at all (Bate 2004, 448–9). But for three years he had lived in the same village with someone who shared his appreciation of 'orchis flowers & fox glove bells', and who in *In Memoriam* would plead with the 'sweet new-year':

> Bring orchis, bring the foxglove spire,
> The little speed well's darling blue,
> Deep tulips dashed with fiery dew,
> Laburnums, dropping-wells of fire.
>
> O thou, new-year, delaying long,
> Delayest the sorrow in my blood,
> That longs to burst a frozen bud
> And flood a fresher throat with song. (Tennyson 1987, 2:395)

Clare yearned for someone to set him free from his imprisonment, and Tennyson for something to purge 'the sorrow in my blood'. It is tempting to imagine that if they *had* met, despite the chasm of class difference that yawned between them, then Tennyson might ultimately have had something more than a failed investment in woodworking machinery to show for his time with Dr. Allen at High Beach. And yet, given the intense suspicion with which poets were regarded by the medical writers of the period, it is not difficult to understand why Allen might have felt that special measures were called for upon discovering that he had *two* such individuals to handle. As Browne and Conolly had emphasised, the exercise of the imagination tended to make poets so vulnerable to insanity that a conscientious mad-doctor could hardly be too careful in dealing with such fragile and mentally unstable souls.

'So Much Has Been Done': The Rise of the British Asylum System

John Clare remained at the High Beach asylum for four years; then, after his famous escape in 1841, he was moved to the Northampton General Lunatic Asylum and remained there until his death in 1864. The Northampton Asylum was a very new institution, which had admitted its first patients in 1838: it was part of the wave of new lunatic asylums which, over the first half of the nineteenth century, were built in almost every county in England, collectively making possible the incarceration of the insane on a historically unprecedented scale (Mellett 1982, 48). In transferring the troubled poet from an old-style private madhouse to a large new public asylum, Clare's family were participating in a great generational shift, which saw care for the insane pass increasingly from untrained private madhouse proprietors to the state-appointed doctors who ran the new asylums. Over the 22 years Clare spent at the General Asylum, it increased hugely in size: originally intended to house just 70 patients when it opened in 1838, by the mid-1860s it held over 400, having effectively become a warehouse for all those residents of Northamptonshire whose families and communities found them too difficult to deal with. (By 1876 even this provision would prove inadequate, leading to the opening of the Northamptonshire County Lunatic Asylum to absorb the overflow [Foss and Trick 1989, 26–8, 93].) This pattern of enormous expansion was mirrored all over the country, as the British government embarked upon an unprecedented, and hugely ambitious, attempt to identify, confine, and institutionalise as many of its insane subjects as possible. As an early patient of the new asylum system, Clare was at the thin end of a very large wedge indeed.

The building of these asylums, at immense cost, bore witness to a new level of concern over the potentially socially disruptive power of madness, and a widespread sense that the time had come for the state to take a much more proactive role in its management. When the first of the new county asylums were built, many contemporary observers thought their size extravagant, and predicted that they would never be more than half-full. It was soon discovered, however, that the asylums filled up almost as fast as they were built, and that no amount of expansion seemed to be able to keep up with the demand for places (Melling and Forsythe 2006, 30; Porter 1987, 166; Scull et al. 1996, 105, 114–7). The level of demand which greeted the opening of the county asylums, which was very much higher than their founders had expected, served to inspire a wide-ranging debate over the nature of insanity: what it was, what caused it, what could cure it, and why there seemed to be so much more of it in Britain than ever before. In retrospect, it seems clear that, as Scull has argued, 'it was the existence and expansion of the asylum system which created the increased demand for its own services, rather than the other way round' (Scull 1979, 245). This, however, was not at all how the matter appeared at the time: and there was much soul-searching regarding the psychic costs of progress, and the mental toll which modern urban life might be taking upon those exposed to it, and much speculation on the role which strong drink, or physical degeneration, or moral wickedness might have played in contributing to the alarming propensity of Victorian Britons to go mad on a seemingly unprecedented scale (Mellett 1982, 63–80; Scull 1979, chap. 7; 1991, 54–5; 1996, 103–4; Sournia 1990, xvii, 23; Rosen 1968, 183–91; Showalter 1987, 24–5; Smith 1999, 108–9; Conolly 1830, 494–5; Shuttleworth 1996, 47; Colley 1983, 12).

One reason for the rise of the Victorian asylum system, then, was a widely felt anxiety over the sheer number of insane people in Britain, and the threat that they might pose to society if left at large. But there was also another, more optimistic, side to the movement: a belief that while madness might be far more common than ever before, it might also be far more treatable than had previously been the case (Shuttleworth 1996, 46). Eighteenth-century asylum-keepers had often seen the role of the madhouse as being simply to provide a secure location in which the insane could be held until it was safe to release them. But the rise of 'moral management' techniques had raised hopes that asylums could become something more than dungeons, and by the 1830s some doctors had begun to argue that asylums could be transformed into spaces of exemplary psychological healthiness, which would do as much to heal the minds of those placed within it as their own dysfunctional environments had done to damage them.

In 1835–1836, Dr. Edward Parker and Dr. Robert Hill began to experiment with implementing a system of 'non-restraint' at the Lincoln Lunatic Asylum, and once their success demonstrated that it was possible to safely run an asylum without keeping the patients constantly locked up or restrained, asylum reform enthusiasts such as John Conolly and W.A.F. Browne began to speculate on just how much more might be achieved if madhouse keepers could be persuaded to renounce the barbaric methods of the past (Jones 1955, 150–1; Scull et al. 1996, 66–71; Colley 1983, 19–20). The enormous sums of money which were extracted from the public purse in order to build the new asylums were taken with the promise that the institutions thus funded would be places in which, with the aid of the latest medical knowledge, the mad could be systematically healed and returned to being productive members of society. In the long run, it was hoped that they might even pay for themselves (Scull 1991, xxii).

Both of these trends were fostered by the developments in medical and psychological writing which, as I have discussed, took place in Britain during the 1820s and 1830s. The conviction that madness might be much more widespread than ever before was partly fostered by the ever-broader definitions of madness which were being used at the time: not just the raving lunacy of the *furiosus*, which even in the eighteenth century would have resulted in the sufferer being locked up, but also forms of 'moral insanity' and *manie sans délire* which, a generation earlier, might have been viewed as wickedness or eccentricity rather than insanity. But as Faas notes, this same body of knowledge, with its confidence that these new forms of madness could be accurately identified and catalogued, was also characterised by the conviction that they could be understood and cured: that the empire of madness, while larger than anyone had previously guessed, might also be more vulnerable to attack than had ever previously been the case (Faas 1988, 9; Colley 1983, 18). The willingness of British asylum-keepers to dispense with restraints after 1835 bears witness to their growing confidence that madness was something which they could reliably control by other means, and that, through their deployment of expert professional knowledge, the insane could be kept at bay without having to rely on the chains and straitjackets of the past.

For doctors such as Browne and Conolly, the late 1830s were thus a time of intense intellectual excitement, in which madness seemed to be well on the way to becoming a solved problem. Scull has ably described the utopian hopes of the early asylum reformers, who believed that their improved understanding of the nature and treatment of insanity might soon permit all but the most stubborn cases to be cured (Scull 1991, xxxv–xxxviii). They were optimistic about the curative powers of the reformed modern asylum, and their optimism made them eager to ensure that everyone who might benefit from a stay in such an institution would actually be placed in one. As the American doctor Isaac Ray wrote in the preface to his 1838 *Treatise on the Medical Jurisprudence of Insanity*:

> While so much has been done, within a comparatively short period, to promote the comfort of the insane, and so much improvement has been effected in the methods of treating their disorder, as to have deprived it of half its terrors, it is both a curious and a melancholy fact, that so little has been accomplished towards regulating their personal and social rights. While nations are vying with one another in the excellence of their public establishments for the accommodation of this unfortunate class of our fellow men, and physicians are every year publishing some instance of an unexampled proportion of cures, we remain perfectly satisfied with the wisdom of our predecessors in every thing relative to their legal relations. (Ray 1838, vii)

Ray, however, was far from 'perfectly satisfied' with the existing laws governing the treatment of the insane, which he saw as far too likely to inflict pointless punishments upon them rather than sending them to be cured at excellent new 'public establishments' such as the county lunatic asylums (Ray 1838, 1–67). It was precisely because he believed these new asylums to be so very effective at curing their patients that Ray thought that the law needed to be reformed; for while it might once have made little difference whether a lunatic was sent to a madhouse or a prison, the two institutions having previously been almost interchangeable, the last few years had seen 'so much improvement' that it was now imperative that such individuals be rescued from the hands of the law and delivered instead to 'public establishments for the accommodation of this unfortunate class', there to benefit from their 'unexampled proportion of cures'. Ray's work towards these ends would ultimately pay off in 1843, when Cockburn drew upon his *Treatise* in his successful legal defence of the paranoid murderer Daniel M'Naghten, in a landmark case which led to a major expansion of the grounds under which defendants in British trials could be found not guilty by reason of insanity—although Ray viewed even these changes as falling far short of the reforms that he desired (Whitlock 1963, 17–18; Ray 1853, 43–53). It was the enormous confidence of doctors such as Browne and Ray in the curative powers of the asylum which helped to persuade reluctant British rate-payers to fund their construction, thus leading to the rise of the new asylum system within which John Clare had the dubious honour of being an early patient.

Another medical writer whose works were influential in the establishment of the 'M'Naghten rules' was Forbes Winslow, who argued in *The Anatomy of Suicide* (1840) and *The Plea of Insanity in Criminal Cases* (1843) that many law-breakers, and all suicides, should properly be understood as victims of mental disease rather than as criminals. Such destructive impulses were, he wrote, 'to a great extent, amenable to those principles which regulate our treatment of ordinary disease', and if those who suffered from them were given proper medical treatment, then enormous amounts of human misery could be averted (Winslow 1840, v; 1842, 178). Winslow's faith in the curative potentials of asylums was such that, in his influential book *On the Preservation of the Health of Body and Mind* (1842), he argued that even individuals who were not yet insane would benefit from institutional medical supervision:

> The Author cannot too often impress upon the attention of his readers, the paramount importance of investigating the morbid manifestations of the mind, with the view of preventing, by an early detection of the premonitory symptoms of Insanity, the full development of the disease. He also feels assured, that were institutions established for the reception and treatment of Insanity in its incipient stage, much benefit would accrue to society generally. (Winslow 1842, vii)

For Winslow, the expanded provision of specialised institutional care did not reduce the need for individuals to engage in continuous self-monitoring of their own 'morbid manifestations of the mind'. Instead, the two went hand in hand, forming complementary parts of the system of internal and external surveillance which was necessary to detect 'insanity in its incipient stage'. The very existence of Winslow's book, which, like John Barlow's near-contemporary *Man's Power over Himself to Prevent or Control Insanity* (1843), was essentially a self-help manual on how to stay sane, bore witness to his belief that insanity was not a rare and exotic condition: instead, it lurked everywhere, 'incipient' in almost every mind, and constant personal and professional vigilance was required to keep it at bay. As Wood writes:

> During the period, the medical profession assumed an unprecedented authority for defining the standard and for predicting the dire consequences of falling outside it. Professional textbooks as well as domestic handbooks pointed to a causal connection between deviation from the prescriptions of self-regulation and a whole range of diseases, from nervous dreads to insanity. (Wood 2001, 64)

Like many other doctors of the time, Winslow blamed gin-drinking and the 'artificial state of society' for this flood of insanity. But he insisted that the battle was not unwinnable, and that the vast majority of cases of insanity could be cured if only they were treated early enough (Winslow 1842, 95–8). It was for this reason that he recommended maintaining ceaseless vigilance of one's own mental state, so that incipient insanity could be identified and treated before it became incurable.

As Winslow made clear in *The Anatomy of Suicide*, this vigilance was especially necessary for 'men of great genius and sensibility':

> It has been observed that the act of suicide may often originate in a feeling analogous to the enthusiasm exhibited by men of great genius and sensibility. This mental idiosyncrasy, which borders so closely on the confines of insanity, has been compared to the narrow bridge of Al Sirat, which leads the followers of Mahomet from earth to heaven, but by so narrow a path that the passenger is in momentary danger of falling into the dismal gulf which yawns beneath him. This abnormal condition of the nervous system is, to a certain extent, dependent on natural organic structure, aided materially by an unhealthy exercise of the imaginative faculty. (Winslow 1840, 121)

Like D'Israeli, Horne, Croker, Madden, Hallam, and Mill before him, Winslow saw 'men of great genius' as possessing an inherently 'abnormal condition of the nervous system', which predisposed them to madness—especially if they overused 'the imaginative faculty'. (In 1842 he would write again of how 'the disposition to *reverie* is very pernicious to intellectual health' [Winslow 1842, 172].) Like Madden, Winslow cited Byron as an example of the proximity of poetic genius to madness, explaining that he had been judged too harshly by the world: Byron, like the criminals and suicides whom Winslow wrote of elsewhere, had been more mad than bad, his bizarre and antisocial behaviour having been the result of epilepsy and mental infirmity rather than voluntary immorality (Winslow 1840, 124–6). Indeed, if an individual with such an 'abnormal condition of the nervous system' did not receive suitable medical support, then crime and suicide could easily be the result:

> An inordinate cultivation of any one faculty of the mind, but more particularly the imagination, will tend to produce the peculiarities which have been illustrated in this chapter. A person who accustoms himself to live in a world created by his own fancy—who surrounds himself with flimsy idealities—will, in the course of time, cease to sympathize with the gross realities of life. The imaginary intelligences which his own morbid mind has called into existence will exercise a terrific influence over him. A German poet commenced writing a poem on the Deity. He allowed his mind to dwell so intensely on the subject, that he fancied he was commanded to "flee from a world of sin and iniquity;" to effect which, he cut his throat, and was found dead in bed, with the razor in one hand and a portion of his poem in the other. (Winslow 1840, 128–9)

Self-murdered by his own over-cultivated imagination, this unfortunate 'German poet' demonstrated all too clearly 'the paramount importance of investigating the morbid manifestations of the mind'. It was in order to prevent such unhappy events that Winslow, like Browne, Ray, and Conolly, called for the reform and expansion of Britain's asylum system, from which 'much benefit would accrue to society generally'—and to would-be criminals, suicides, and geniuses, in particular (Winslow 1842, vii).

For Tennyson, the immense expansion of the physical and legal infrastructure for the control of the insane in Britain posed little in the way of direct threat. As a gentleman, he was at scant risk of incarceration unless his behaviour became very erratic indeed, and even then he would have been unlikely to end up at a county asylum: at worst, he would have been placed in a private madhouse, like his unfortunate brother Edward. Safe though he may have been, however, his awareness of the ongoing campaign to identify and institutionalise the entire mad population of Britain may have lent an additional sense of urgency to his own struggles to distinguish between sanity and madness. As Faas writes, the widening definitions of insanity used in the period 'must have been a source of genuine apprehension' to 'the neurotic and eccentric', and Tennyson was both (45–6). Matthew Allen, meanwhile, was an enthusiastic believer in the curative powers of the reformed asylum, and had been an early adopter of

Parker and Hill's system of non-restraint. But despite all Allen's modern methods, Clare's condition stubbornly failed to improve during the time he spent at Allen's asylum: and while Tennyson struggled to construct a form of post-Romantic poetry which might pass the test of sanity, one of the last of the great British Romantics was sliding away into permanent madness almost on his very doorstep.

'The Fields Were Sudden Bare': Enclosure and Insanity in the Poems of John Clare

Tempting though it is to see the county asylum system as simply another brick in the wall of coercive social oppression which so enraged John Clare, there is little evidence that it was regarded in this way at the time. The motives for the construction of the county asylums were partly pragmatic and partly humanitarian: humanitarian insofar as they resulted from moral outrage at the scandalous treatment of the insane in workhouses and private madhouses, and pragmatic insofar as the insane were seen as a drain on the resources of the nation, making it desirable to contain them in order to prevent them from doing further damage, and, when possible, to cure them so that they might resume lives of productive labour (Scull 1991, xxii). The eagerness with which poor families sought to take advantage of the services offered by the asylum system demonstrates that it was widely seen as a useful public resource rather than as an instrument of state tyranny, and the willingness of such families to send insane relatives to be cared for at the county asylums forms a striking contrast to the lengths which they were willing to go to in order to preserve their loved ones from having to enter workhouses, which remained widely hated (Smith 2006, 116–17; Scull 2006, 77). Nonetheless, in a broader sense, both workhouses and asylums clearly did form elements of what Foucault famously called 'the great confinement': the increasingly ambitious attempt of European states to exert an unprecedented level of control over their own people (Foucault 2006, 66).

For Clare, a third element in this system of control was the ongoing enclosure of common lands in England, the progress of which he bitterly and eloquently lamented in his poetry (Cronin 2002, 205–6). As well as being a space for communal grazing of animals and gathering of firewood, the old village common had also been the traditional haunt of the homeless pauper and the wandering lunatic: but now the pauper was to be sent to the workhouse, the lunatic to the asylum, and the common itself carved up by local landowners, transformed into a private space owned by specific individuals (Hayes 1987, 10–19). The parallel is underlined by the fact that the Second General Inclosure Act was passed in 1845, in the same year as the Second Lunacy Act: both were part of the general legislative tidying up of Victorian Britain, which aimed to organise, rationalise, and render productive a greater proportion of the land and population than ever before. The logic of such legislation was that the new agricultural system, supported by enclosure and the New Poor Law, represented a triumph of rationality and sanity over muddle and inefficiency. The enclosed landscape was a sane landscape, which had been rendered as efficiently productive as possible, just as the workhouse and asylum systems were supposed to increase the sanity and productivity of the population which inhabited that landscape, doing away with madness, idleness, and waste.

As Scull notes, the mad were distinguished from other able-bodied deviants precisely by the fact that they could not be compelled to labour. The madman was thus the human equivalent of the unproductive wasteland, awaiting the intervention of the 'improving' mad-doctor to render him economically productive once more (Scull 1979, 132–41). This may have been how the enclosure system appeared from above, but it did not always appear so from below: to the inhabitants of John Clare's native Northamptonshire, for example, which had a tradition of popular resistance to enclosures running all the way from the Midland Revolt of 1607 to the Swing Riots of 1830. For Clare, 'enclosure' was simply a process through which 'The common health, became the spoilers prey', leading to the replacement of 'comforts cottage' with 'workhouse prisons' built to accommodate all those whom the enclosure acts had driven from the land (Clare 2004, 98).

Long before his consignment to High Beach asylum, Clare had drawn explicit parallels in his poetry between the agricultural enclosures around him and his own state of mental desolation. For Clare, the enclosed, managed landscape was one which had been robbed of all those comforts and graces which made life liveable, leaving only a desert of uniformity maintained through cruelty and greed. Its mental equivalent was not a state of sanity, but a state of misery or depression: an enclosed field was like a man in despair. This equivalency is made very clear in his early poem 'The Lamentation of Round-Oak Waters' (1818), in which he writes:

Oppress'd wi' grief a double share.
  Where Round oak waters flow.
I one day took a sitting there.
  Recounting many a woe.
My naked seat without a shade.
  Did cold and blealy shine.
Which fate was more agreeable made.
  As sympathising mine. (Clare 2004, 18)

The 'naked' landscape mirrors the desolation of Clare's mind, 'sympathising' with his fate—and well it might, for it turns out that their woes have the same source. Rising from the river, the spirit of the brook spells out to Clare the similarity of their sufferings:

'Ah luckless youth to sorrow born
  Shun'd Son of Poverty
The worlds made gamely sport and scorn
  And grinning infamy
Unequall'd tho thy sorrows seem
  And great indeed they are
O hear my sorrows for my stream
  You'll find an equal there. (Clare 2004, 19)

The equal 'sorrows' of the spirit are due to the enclosures which have been inflicted upon the landscape around it, turning it into a sort of physical reflection of Clare's mental suffering. The parallel between what the rich have done to Clare and what the rich have done to Round-Oak Waters is underlined by the fact that the same verb, 'cutting', is used to describe both: their cruelty to him is described as being 'more cutting than the storm', while the river-spirit complains that 'the moors / Are all cut up and done', and that the men who felled his willows were 'hard as the ax that Cut them down'. 'These all our Ru[i]n planned', the spirit tells Clare, referring to both the physical ruin of the enclosed rural landscape and the emotional ruin of Clare's peace of mind (Clare 2004, 19, 21, 23). Both are the result of greed and cruelty, and each is an 'equal' symbolic reflection of the other.

In his later poem 'Remembrances', written in the mid-1830s, Clare once again drew a direct parallel between the physical transformation which had passed over the landscape and the mental transformation which had passed over him during the same years:

O I never thought that joys would run away from boys
Or that boys would change their minds and forsake such summer joys
But alack I never dreamed that the world had other toys
To petrify first feelings like the fable into stone
Till I found the pleasure past and a winter come at last
Then the fields were sudden bare and the sky got overcast
And boyhoods pleasing haunts like a blossom in the blast
Was shrivelled to a withered weed and trampled down and done
Till vanished was the morning spring and set that summer sun
And winter fought her battle strife and won. (Clare 2004, 260)

Clare's phrasing effectively dissolves the boundaries between his own mind and the enclosed landscape: when he says that 'the fields were sudden bare', it is simultaneously a literal description of what has happened to a place he used to play in since its enclosure and a metaphorical description of what has happened to his emotional state since the days when he used to play there. In 'The Lamentation of Round-Oak Waters', the enclosure of the landscape is described as being directly comparable to the desolation of Clare's mind, and caused by the same forces, but in 'Remembrances', they seem almost to have collapsed into a single event, making it impossible to say whether the enclosures have caused his depression, or whether they simply happened to advance in parallel.

The enclosed landscape looks like this to Clare because he is depressed, but also because it is depressing. Far from encouraging a state of healthy, productive rationality, it fills him with a sense of helpless misery, driving him towards mental collapse. His lament takes in the 'poor swains' who are forced to first create, and then inhabit, this desolated landscape, but it is also directed towards the landscape itself, which the enclosures have stripped of birds and animals, trees and flowers and streams—the very things which Clare, like Blake and Wordsworth before him, had found to be the fittest subjects for his poetry. In removing them from the landscape, the enclosures have robbed the landscape itself of its aesthetic and spiritual meaning. Enclosed, it ceases to be a fit subject for any form of poetry except the lament.

Just as Clare felt that enclosure had robbed the Northamptonshire landscape of its identity, so he felt that his own identity was lost over the course of his own long enclosure—first within Allen's private madhouse and then within the Northampton County Asylum, where, as he famously lamented, 'I am: yet what I am none cares or knows' (Clare 2004, 361).

His poems emphasise the extent to which these systems of control, which were intended by their architects to restore order and reason, could in fact be experienced as alienating and disorientating by the people actually subjected to them. It was just such a group of people, enraged and bewildered by the New Poor Law and the enclosure acts, to whom another early graduate of the county asylum system, John Tom, made his appeal in 1837–1838 (Reay 1990, 77). The fact that both Tom's actions and the actions of his followers were swiftly categorised as 'madness' demonstrates the extent to which, in this period, any kind of action which was disruptive of the new social order was liable to be dismissed as insanity, whether it was carried out by revolutionaries or would-be Romantic poets.

Romantic Madness: The Career of John Tom

It is often asserted that the 1746 Battle of Culloden was the last battle fought in mainland Britain. A less well-known contender would be the Battle of Fishguard, 1797, in which a French invasion force landed in Wales, skirmished with local militias, and then surrendered after its commander realised that his position was untenable. But there is a third possibility: the Battle of Bossenden Wood, fought on 31 May 1838. There, in a Kentish forest, Sir William Percy Honeywood Courtenay, Knight of Malta, rightful heir to the county of Devon, and self-proclaimed messiah, made his last stand against Queen Victoria's soldiers, and there he died in battle, along with eight of his followers and two of the Queen's men. Courtenay's doomed rebellion may have marked the end of the centuries-long struggle of the British state to secure a monopoly on large-scale organised violence within its own borders, but it was also an episode in another long historical struggle: the war on madness, which in 1838 was only just beginning in earnest. For the man who called himself Sir William Courtenay was, in fact, neither a knight of Malta nor the heir to the county of Devon. His real name was John Tom, and he was a maltster from Cornwall, who had been recently released from the Kent County Lunatic Asylum (Rogers 1962, chaps. 4–6; Bloom 2013, 70–4; Reay 1990, 110–16). Like the Northampton Asylum that Clare was sent to in 1841, the Kent County Lunatic Asylum was an institution of very recent foundation, having first opened in 1833: Tom had been one of its first inmates. He spent four years there, and he appears to have been a model patient, but when he left he refused to return to the 'sane' life in Cornwall which he had forsaken (Rogers 1962, chap. 4; Reay 1990, 116). Instead, he took up arms, Quixote-like, against what the world called sanity: and in the end, confronted by the Queen's soldiers, he evidently preferred to kill and die as Sir William Courtenay than to live and let live as plain John Tom of Truro.

I would suggest that Clare and Tom are linked by more than just chance similarities in dates. Faced with financial failure and the encroachment of depression-inducing meaninglessness, Clare and Tom both took on 'romantic' aristocratic fantasy personae: Tom claimed to be first Count Rothschild and then Sir William Courtenay, while Clare began to assert that he was in fact the deceased Lord Byron, even though he had observed Byron's funeral procession with his own eyes (Halmi 2004, 144). They made aggressive but futile gestures of universal defiance: Tom in his doomed rebellion, Clare in the challenges he issued from Northampton Asylum, insisting that he was a prize-fighter who had never been defeated and daring anyone in the world to come and face him in a boxing match (Faubert 2005, 286). When Tom took up arms as Sir William, he rallied the men of Kent to his standard with a manifesto which was, in every sense of the word, romantic; and, like Clare, he seems to have made heavy use of the works of Byron in constructing his fantasy persona (Reay 1990, 126–7). He reinvented himself as an aristocrat denied his rightful inheritance, come to lead the people against the injustices of social inequality: a figure not dissimilar to the hero of Byron's *Lara*, or indeed to Byron himself.

Tom employed, in his own idiosyncratic way, the same combination of anachronistic and exotic 'romantic' fantasy and utopian radical politics which had characterised the works of the major Romantics a generation before. He shared their concern for the welfare of the rural poor, and the men whom he led to Bossenden Wood were mostly agricultural labourers: men of the same class that John Clare had been born into, and whose lives and labour Clare, like Burns and Wordsworth before him, had chronicled in his poetry (Rogers 1962, chaps. 2, 5, and 6). He claimed to have travelled in the East, like the protagonists of Byron's oriental tales, and to have learned the ancient secrets of Egypt and Syria (*Morning Chronicle*, 2 June 1838). In the claims of both Tom and Clare, there was much which was reminiscent of the British Romanticism of the previous decades, including exaltation of imagination, defiance of mainstream political ideology, reliance on personal inspiration, and concern for the working poor. But if this was Romanticism, then it was a Romanticism so extreme that those around them regarded it as madness. Both men disowned their 'real' lives (and their real wives), retreating from their disappointments into ever-more extravagant forms of fantasy.

While Tennyson and Clare were settling into their respective lodgings at High Beach, 40 miles to the south-east John Tom was gathering followers among the people of Kent. After his death, the journalists who wrote about his activities asserted that he had clearly been insane, and over a century later Tom's biographer, P.G. Rogers, agreed with this judgement, confidently diagnosing him as 'a paranoiac' afflicted with 'delusions of grandeur'. As Rogers himself noted, however, paranoia is 'the sanest form of insanity', and a later historian, Barry Reay, has questioned whether Tom was mad at all (Rogers 1962, 208; Reay 1990, 122–4). Tom certainly did some extremely odd things during the 1830s: he abandoned his family, told a series of extravagant lies about his identity, claimed to possess supernatural powers, and finally murdered a man with extreme violence. But in other respects, Tom did not at all resemble the traditional madman as he tended to be depicted in the art and literature of the period: he never raved or spoke in nonsense, never claimed to have heard voices or suffered from hallucinations, never harmed himself or attacked other people without warning, never tore his clothes or neglected his personal hygiene, and so on. Clearly, he was an eccentric, arrogant, self-obsessed fantasist. But was he actually insane?

Ultimately, the most significant fact about Tom's extraordinary masquerade—and the one which proved most disturbing and embarrassing to the authorities after his death—was that it *worked*. In a semi-fictional biography which appeared shortly after his death, 'Canterburiensis' remarked:

> One of the greatest and most extraordinary features in the history of the maniac Courtenay, for we know not how to distinguish him by any other epithet, will be found in the unprecedented influence, which he obtained, and permanently exercised over a large number, not only of the agricultural population, but of the voters of the city of Canterbury[.] (Canterburiensis 1838, iv–v)

Tom did not spend the years 1832–1838 raving in a gutter, the helpless prisoner of his own deranged beliefs, like the archetypal 'mad Tom' of literary legend Instead, he gathered followers, ran for parliament, made speeches, started a newspaper, and finally assembled a band of devotees who proved willing to follow him to their deaths. Given his obvious desire for attention, his assumption of a series of fantasy personae served clear and logical ends, and Reay has argued that his performances need to be understood within the context of contemporary charismatic radicalism rather than as straightforward symptoms of mental illness (Reay 1990, 116–20). 'Canterburiensis' attributed to him the belief 'that he was destined to achieve some great purpose in the world' (Canterburiensis 1838, 153). If Tom was mad, then his madness may have been simply the not uncommon 'madness' of feeling that he deserved to be recognised as special despite his apparent ordinariness—a trait which he shared with many of the would-be poets of the era—rather than the more exotic madness of really believing, despite the considerable evidence to the contrary, that he was actually a Knight of Malta.

Unlike John Clare, who really does seem to have suffered acutely from paranoid delusions, Tom was consigned to Kent asylum more because his behaviour was becoming troublesome than because he was obviously drifting out of touch with reality. As his condition worsened, Clare began behaving in a range of alarming ways: he saw evil spirits, refused to walk down a street because he was convinced that the devil was waiting there for him, yelled accusations at the actor playing Shylock during a performance of *The Merchant of Venice*, and claimed to be married to his childhood sweetheart, Mary, as well as his actual wife, Martha (Blackmore 1986, 211; Faubert 2005, 278; Foss and Trick 1989, 126–7; Bate 2004, 335–6, 398). Tom never did anything like this: he was simply locked up because his actions were so odd and socially disruptive that the madhouse was felt to be the appropriate place for him. The alternative would have been transportation—and had his trial taken place a few decades earlier, when a narrower understanding of insanity still prevailed, transportation would probably have been his fate (Rogers 1962, 67–71).

Tom's confinement was thus symptomatic of the general shift in the purpose of British asylums over the first half of the nineteenth century, from places meant primarily for the imprisonment of the hopelessly crazed to institutions for the containment of a much wider range of disruptive social misfits. As I have discussed, this shift was facilitated by the rapidly widening definitions of insanity which were starting to be used at the time, so that an ever larger range of the unconventional, eccentric, and just plain odd found themselves reclassified as insane (Reay 1990, 123). As religious excess was viewed as a standard cause of madness, people who professed unconventional religious beliefs were particularly likely to find themselves categorised as lunatics: it was on these grounds that the street preachers Jacob Scott and Mary White were institutionalised in 1828 for declaring themselves to be prophets, while the unfortunate George Davenport was declared insane in 1837, despite seeming quite sane when examined, simply because he insisted on giving away what were regarded as excessive sums of money to religious causes (Smith 1999, 101–10; Suzuki 2006, 142–3). When Carlyle visited the Irvingites in 1831, he confidently declared that those of them who spoke in tongues were 'crazed', and dismissed the one woman whose glossolalia he actually heard as a 'hysterical madwoman' who would be better served by having cold water thrown over her than by actually being listened to (Pearsall 2008, 64, 67).

The contemporary newspapers which reported on the Battle of Bossenden Wood tended to take Tom's insanity as given, generally treating the fact that he had spent time in a lunatic asylum as irrefutable proof that he was, in fact, insane (Reay 1990, 125; *The Times*, 1 June 1838; *Morning Chronicle*, 2 June 1838; *Manchester Times and Gazette*, 9 June 1838). His followers, too, were described as lunatics: thus, when the correspondent from the *Morning Chronicle* discovered that the local peasantry regarded Tom and his dead followers as martyrs, and that some of them expected Tom to rise from the dead, he declared that 'incredible as it may appear, I have been assured of this as a positive fact with respect to the utter folly and madness of the lower orders here' (*Morning Chronicle*, 2 June 1838). Commenting on the willingness of Tom's followers to attack the soldiers sent to subdue them, even though they were outnumbered and mostly armed only with clubs, he remarked that 'nothing but the madness and fanaticism that possessed them can account for such conduct'; similarly, the correspondent for the *Brighton Patriot* asserted that 'the madmen who followed [Tom] were eager to fight the military' (*Morning Chronicle*, 2 June 1838; *Brighton Patriot*, 5 June 1838).

In these reports the word 'madness' is clearly being used in a very general way, in order to account for all those aspects of the situation which the correspondents found most difficult to understand. Why did John Tom claim to be William Courtenay? Madness. Why did people believe him and follow him? Madness. Why did he lead them in armed revolt against the government? Madness. Why did they attack the Queen's soldiers when he told them to, despite being outnumbered and outgunned? Madness. 'Madness', in other words, turns out to be a catch-all term for any kind of surprising behaviour, which pre-empts the need for any kind of explanation by asserting that there is nothing to explain, reducing these disturbing events to the irrational actions of lunatics. Yet even after asserting that both Tom and his followers must have been insane, the *Morning Chronicle's* correspondent was clearly still uneasy about the whole situation. He described Tom's influence as 'most extraordinary', and stated that 'strange as it may appear, this madman obtained influence over the deluded people, and the result is now seen' (*Morning Chronicle*, 2 June 1838). His choice of words highlights the inadequacy of the explanation offered: for it does indeed seem 'strange' and 'most extraordinary' that someone so obviously mad should have attained such 'influence', and that all his followers should also have been madmen seems more strange and extraordinary still.

The *Chronicle*'s correspondent described the people of Kent as 'deluded' for believing in Tom. In doing so, he may well have had in mind a fragment of verse written by one of those very believers, which asserted precisely the opposite:

> Is it a delusion? No, its peace I hear.
> As yet welcome sweet guest.
> A passing spirit softly wispers [sic].
> Him safe from harm—and when.
> The loud clash of war's alarm attacks.
> Him and boasts the tyrants proudly.
> Round him still his manly heart.
> Shall know no fear—.

These verses, as the *Chronicle* and several other newspapers reported, were found written on a slip of paper in Tom's pocket after his death in Bossenden Wood. The verses were written in 'the handwriting of a female', the paper on which she had written them had been perforated by a musket ball, and was soaked in Tom's blood (*Morning Chronicle*, 2 June 1838; *Manchester Times and Gazette*, 9 June 1838; *York Herald*, 9 June 1838—each reprints the poem with minor variations in orthography.) Their author, clearly, knew exactly how her belief in Tom's claims would be interpreted by the outside world: thus her very first concern is to insist that her belief is *not* 'a delusion', and is in fact founded upon personal supernatural revelation, namely, the 'wispers' of 'A passing spirit' which she implies that she has personally heard. These lines may have been written by Sarah Culver, a follower of Tom whose faith in him was so complete that she was still insisting that he had been the second coming of Christ several days after he met his death in battle. Upon her, the *Morning Chronicle's* correspondent passed a predictable judgement: 'It cannot be doubted that her intellects are deranged' (*Morning Chronicle*, 5 June 1838).

As the aftermath of Tom's rising demonstrated, the newly expansive definitions of madness which came into use during the 1830s could be used to de-legitimise entire movements, whose members could be dismissed en masse as lunatics simply on the grounds that no sane person could possibly hold the political or religious beliefs that they espoused. (Browning played upon these ideas in his 1841 poem *Pippa Passes*, in which Luigi, who confesses that 'I think my mind is touched', struggles to work out whether his eagerness to martyr himself for the cause of Italy's freedom is simply a side effect of his unbalanced mental condition [Browning 1983–2009, 3:68].) As long as madness had been seen as a rare and extreme state, then it could potentially be invoked to explain the bizarre actions of unhinged individuals, but it could hardly account for the beliefs and actions of entire groups or populations. Revolutions were thought to cause madness, but not to be caused by it: and eighteenth-century thinkers had preferred to explain such events in terms of 'enthusiasm', 'fanaticism', or 'superstition', instead, all of which implied that most of the people involved were the victims of deception or manipulation rather than actual lunatics (Rosen 1968, 179–84; Scull 1991, 28–9, 61–4; Shuttleworth 1996, 48).

As I have discussed, however, by the later 1830s many people in Britain had come to accept that madness was (or perhaps *had become*) much more prevalent than their ancestors had ever imagined. This belief that madness was not rare but commonplace made it available as a possible explanation for a much larger range of human behaviours, including drunkenness, fringe religion, criminality, and even political radicalism. Carlyle, in his influential 1837 history of the French Revolution, defined the revolution in the following terms:

> *La Révolution* is but so many Alphabetic Letters; a thing nowhere to be laid hands on, to be clapt under lock and key: where is it? what is it? It is the Madness that dwells in the hearts of men. (Carlyle 1837, 3:345)

For Carlyle, who was fond of apocalyptic sublimes, outbreaks of mass insanity like the ones which he saw as having triggered the French Revolution were by no means always something to be disapproved of. Describing the events of August 1792, he asserted that 'had *not* that same great Nether Deep, of Bedlam, Fanaticism and Popular wrath and madness, risen unfathomable on the Tenth of August', then 'French Patriotism' would have been reduced to 'an eloquent Reminiscence; swinging on

Prussian gibbets' (Carlyle 1837, 3:122). Few other writers at the time viewed the revolutionary 'Madness that dwells in the hearts of men' with equal tolerance, however; and when contemporary political or religious groups such as John Tom's followers (or, indeed, the Swing rioters, Chartists, or 1848 revolutionaries) were described in such terms, it was usually as a straightforward way of dismissing the validity of their beliefs, and asserting the need for their containment (Rosen 1968, 181–2). Sournia notes that 'since the proletariat, drunkenness and revolution were seen to be inextricably linked, the working classes had to be controlled in the interests of society as a whole', and as drunkenness and revolution were seen as being both causes and symptoms of what was believed to be an epidemic of insanity among the lower classes, it was only natural for doctors and asylums to take on a significant role in this process of social control (Sournia 1990, 103–4). Thus Rylance has called attention to the fact that the asylum-keeper J.G. Millingen published two related books, *Recollections of Republican France* and *The Passions*, in the revolutionary year of 1848. For Millingen, sanity was defined by the maintenance of proper control over the passions: the French Revolution demonstrated what happened when this control was neglected on a nationwide scale, resulting in a universal epidemic of insanity, and asylums such as his own had a crucial role to play in preventing Britain from falling into the same unhappy state (Rylance 2000, 116–18, 126).

An unusually systematic version of this theory appeared in 1857 in a book entitled *Phantasmata*, which was written by none other than the author of *The Infirmities of Genius*, Richard Madden. Over the course of this lengthy work, Madden attempted to explain a whole variety of extraordinary historical events, including Tom's rebellion, in terms of 'epidemic mental disorders'. Madden insisted that this attempt to explain large sections of history in terms of contagious mental pathology was not reductive; on the contrary, it was those who tried to explain 'remarkable outbreaks of popular phrenzy' in purely contingent historical terms, rather than as medical phenomena, whom he regarded as people trying to save themselves from 'the trouble of thinking too deeply' (Madden 1857, 1:vi–viii). Revolutions thus joined Romantic poets on the list of things which Madden viewed doctors as being uniquely well-qualified to understand and explain. Indeed, having explained the French Revolution to his own satisfaction, Madden then used *Phantasmata* to reiterate his view that the careers of poets such as Blake and Shelley (both of whom he mentioned by name), as well as those of mystics such as Swedenborg, were best understood as products of their extremely delicate nervous systems and over-active imaginations (Madden 1857, 1:90–3, 149–50). At the very historical moment when Burns, Shelley, and Byron were starting to serve as both poetical and ideological models for a new generation of politically radical Chartist poets, some of whom even indicated their radical commitments by getting Byronic haircuts, radical politics thus threatened to join excessive imagination and unconventional religious beliefs on the lengthening list of traditionally Romantic qualities which were coming to be not only associated with madness, but actually interpreted as forms of insanity in their own right. (Janowitz 1998, 148–51).

'Secrets of the Brain': Poetry, Psychology, and the Art of Staying Sane

In 1867, 30 years after Tom's death, Dickens imagined how Tom might have felt during his years in Kent asylum:

> Yes! Yes! There in the sun stands the angel of Revelations with the book in his hand, and the voice comes proclaiming John Tom a prophet of the Lord, sent through the world to lower the price of bread, and to work wonders, by fire and the sword, until the unbelieving turn to him, and acknowledge him as the chosen of Heaven.
>
> Nothing can shake this man now. He is the chosen prophet, and when the angels touch the doors they will open. Four years he waits, and one morning the doors open according to the prophecy. Madness is cunning. Tom is silent about his mission of fire and sword; he is sane on some topics. He sees the angels leading him from the accursed place, but he does not speak of them, lest the madmen should thrust him back again. His madness is no longer secret and mistrustful; it has risen to delirium, and his brain is on fire at the rapid approach of his avatar. Voices assure him from the great sunset clouds, red with the coming vengeance; voices rise from the hop-fields, from the flowers; the very birds proclaim him prophet and chosen; the brooks ripple out their welcome, the great angel in the sun waves a sword, every sunbeam leads him to victory and glory [...]. (Rogers 1962, 73)

This is fine stuff, but there is little reason for believing it to be accurate. As Rogers notes, Dickens' description ignored Tom's 'exemplary conduct, and his sustained effort to prove that he was perfectly normal' (Rogers 1962, 74). Tom never claimed to hear voices, to see visions, or to communicate with angels, and the passage about hearing voices in the clouds, the fields, the flowers, the brooks, the birds, and the sun does not sound like Tom at all, either in the poems and articles he wrote before his confinement to the Kent asylum or in the speeches recorded by his doctors during it ('Canterburiensis' 1838, 274–337; Reay 1990, 122–3). What it *does* sound like is Romantic poetry. Tom may never have said that the clouds, birds, fields or flowers spoke to him, but Blake, Wordsworth, Shelley, and Clare made such claims all the time. As Colville notes, 'English Romantic poetry has *at its core* the experience of vision, and the values enforced by that' (Colville 1970, 2). Blake even stated explicitly that true 'visionary perception' consisted precisely in seeing the sun, for example, in almost the same way that Dickens imagined it appearing to John Tom in his madness:

> It will be Questiond When the Sun rises do you not see a round Disk of fire somewhat like a Guinea O no no I see an Innumerable company of the Heavenly host crying Holy Holy Holy is the Lord God Almighty. (Blake 1988, 565–6)

Blake, of course, had been extreme even by Romantic standards. Regularly accused of insanity throughout his career, he struck back by insisting that 'the enjoyments of Genius' look like 'torment and insanity' to those who cannot understand them, and that a dose of what the world called madness could be very advantageous to a poet, mischievously claiming that the problem with his deeply troubled contemporary William Cowper was that he had not been nearly mad enough (Blake 1988, 35, 663). But the idea that one of the crucial qualifications for being a true poet was the ability to see the natural world as it appears when transfigured by the power of imagination—to see the angels in the sun, to hear the voices rising from the hop-fields—appear in the works of a number of other Romantic writers. Wordsworth, for example, wrote in *The Prelude* of hearing 'The rocks that muttered close upon our ears', and 'Black drizzling crags that spake by the way-side / As if a voice were in them', both of which he took to be 'Characters of the great Apocalypse / The types and symbols of Eternity' (Wordsworth 1985, 131). In his 1821 poem 'The Fate of Genius', Clare had described a man who was regarded as 'crazy' by his community, because he talked to the flowers 'in visions': and yet this individual is not a lunatic but a 'rustic genius', clearly modelled on Clare himself (Clare 2004, 80–3). The fact that Dickens used such rhetoric to describe a state of madness rather than a state of visionary inspiration, and made Tom fit in with his preconceived ideas of how a madman might think by making him sound like a Romantic poet, suggests yet again how easily such states could appear interchangeable, and how carefully poets such as Tennyson needed to write in order to avoid giving an impression of 'delirium'.

After taking one of his regular walks outside the asylum, John Clare wrote a description of the dew on the grass:

> The dew drops on every blade of grass are so much like silver drops that I am obliged to stoop down as I walk to see if they are pearls, and those sprinkled on the Ivy woven beds of Primroses underneath the hazels, white thorns and Maples are so like gold beads that I stooped down to feel if they were hard but they melted from my finger. (Clare 2004, 413)

Is this poetic imagination, or is it madness? Clare seems to be doing exactly what all the great Romantics recommended: he is breaking free from the shackles of conventional perception, which insists that dewy grass is just so much soggy vegetable matter, and seeing instead gold, silver, pearls, a visionary universe of beauty, so vivid that he has to touch the water droplets to check that they are not really pearls. But less sympathetic observers, like the villagers who dismiss the 'rustic genius' as a 'crazy man', or the Somersby villagers who thought Tennyson was 'craäzed', might well feel that a man who has to reach down to touch the grass in order to be sure that the dew drops are actually made of water could not be entirely right in the head (Tennyson and Dyson 1974, 115–16). From such a perspective, Clare's vision of the pearls on the grass would simply confirm the truth of Conolly's warning that a poet who cultivated his imagination to the point where 'He hears the voice of God in the wind', while neglecting the 'comparing faculty' (which would have told them that the dew drops couldn't *really* be pearls) was at risk of sliding into insanity—and that, having reached such a state, a room in one of England's new asylums might really be the best place for him (Conolly 1830, 81–4, 131, 222–3).

Unlike John Clare (or John Tom), Tennyson was never threatened with involuntary confinement, and he seems to have viewed the advance of contemporary psychology as one of many regions in which the new scientific knowledge of the 1830s was making inroads into the ignorance of the past. In the 'Envoi' to his poem 'The Day-Dream', probably written in 1837–1838, he invited his addressee to imagine the benefits of sleeping for a 100 years:

To sleep through terms of mighty wars,
  And wake on science grown to more,
On secrets of the brain, the stars,
  As wild as aught of fairy lore;
And all that else the years will show,
  The Poet-forms of stronger hours. (Tennyson 1987, 2:48, 2:57)

Here, the 'secrets of the brain' which the 'science' of future generations will discover constitutes a new mythology in its own right, just as 'wild' as the 'fairy lore' which it will replace, and presumably just as capable of inspiring the 'Poet-forms' of the future. At the very least, the fact that this future age still *has* 'Poet-forms' suggests that knowing the 'secrets of the brain' is not incompatible with poetry, and might even be able to enhance it—perhaps, as Madden had argued, by helping to preserve the sanity of poets.

One example of the ways in which the newly discovered 'secrets of the brain' might be used to maintain one's sanity, just as Winslow and Barlow recommended a few years later, can be seen in the stanzas which, in 1837, Tennyson added to his 1834 poem 'Oh! That 'Twere Possible'. In the original version of this poem, which Tennyson wrote during the period of deep grief which followed Hallam's death, he described a man being haunted by the 'abiding phantom cold' of his dead beloved, a shadow that 'flits before me—/ Not thou, but like to thee' (Tennyson 1987, 2:20–22). In the 1834 text of 'Oh! That 'Twere Possible' this haunting was presented as permanent and hopeless, leaving the speaker crippled with misery, and the poem ended with an expression of utter despair:

Always I long to creep
To some still cavern deep,
And to weep, and weep and weep
  My whole soul out to thee. (Tennyson 1987, 2:22)

In 1837, however, Tennyson added a new set of stanzas to the poem, in which its speaker attempted to rid himself of this persistent ghost and the misery that it brought with it:

Get thee hence, nor come again,
  Pass and cease to move about—
Pass, thou death-like type of pain,
  Mix not memory with doubt.
'Tis the blot upon the brain
  That *will* show itself without. (Tennyson 1987, 2:20–3)

Within the poem, this comes as a very sharp change of direction. The 1834 text gave no indication that the speaker knew or cared what the true nature of the 'phantom' was, being concerned only with its emotional effects upon him. Here, however, it seems that the speaker has read Scott's *Letters upon Demonology*, and decided to put their teachings into practice. Like the user of Conolly's 'comparing faculty', he refuses to conclude that the fact he sees a ghost everywhere he goes means that the ghost is actually real; instead, he reasons that he is much more likely to be suffering from hallucinations as a result of some kind of physical malfunction. A few lines later, he asserts that the 'phantom' he sees is merely 'a dull mechanic ghost / And a juggle of the brain': 'mechanic' in the sense that its appearance is simply a mechanical consequence of the 'blot' on his brain, and as such has no spiritual meaning or significance.

Having concluded that the ghost he sees doesn't really exist, Tennyson's speaker makes up his mind to ignore it, in the hope that this will make it go away. When this doesn't work, he tries to conjure a kind of 'anti-ghost' of his own, by deliberately calling to mind the image of his dead beloved, contrasting this voluntarily imagined 'phantom of the mind' with the 'ghastly sister' which haunts him. *His* phantom, the speaker asserts, is 'a phantom fair and good', because it comes and goes when he desires it to, unlike the 'ghastly sister' which is animated by 'the moving of the blood / That is moved not of the will'. The poem thus distinguishes firmly between two versions of imagination, one good and one bad. The good kind of imagination is voluntary, and under the command of the 'will'; the bad kind is involuntary, a product of the 'brain', 'blood', 'nerves', and 'eye'. The bad imagination is bodily, and will die with the body; but the good imagination is spiritual, and will lead the speaker 'like a star' (shades of *Adonaïs*) to the celestial realm where the *true* spirit of his beloved awaits him (Tennyson 1987, 2:22–3). It is because the speaker is an educated man, one who knows about the 'secrets of the brain' and the ways in which the mind can be affected by 'the moving of the blood', that he does not, like one of Scott's superstitious savages, attach a significance to his physically induced hallucination which it does not deserve. There is no indication that he has succeeded in getting rid of it—indeed, he seems resigned to having it follow him around for the rest of his life—but seeing it no longer makes him want to creep away into a cave and die.

The 1837 version of the poem thus implicitly offers a potential solution to the problem of how to distinguish between poetic imagination and madness. If the objects of imagination are called up deliberately and voluntarily, in the way that the speaker calls up his 'phantom of the mind', then they are compatible with sanity; but if they arise involuntarily, then they may be the result of 'blot[s] upon the brain', and must be systematically ignored and disbelieved if madness is to be avoided. Judged by this standard, Clare, like Browning's Aprile, could only be found wanting, as his vision of the dew drops as pearls did not come and go at the command of his 'will': instead, it persisted so powerfully that he found himself 'obliged to stoop down as I walk'. Colley has argued that it was precisely because Tennyson adopted this kind of 'clinical' approach to his own abnormal mental states that he was able to avoid following Clare into madness (Colley 1983, 4–8). And yet Tennyson's model of non-pathological imagination is, in its own way, just as much of a 'mechanic ghost' as the pathological, involuntary hallucination which it is supposed to upstage. The speaker imagines the dead woman, and she appears; he calls her to his side, and she comes. She may be a 'phantom fair and good' but she is still a mere marionette, possessing none of the uncanny life of her 'ghastly sister'.

This mental puppetry, in which the imagination simply calls known figures from the memory and moves them around on pre-determined paths, is a far cry from the encounters with angels that Tennyson described in 'Armageddon', 'Timbuctoo', 'The Mystic', and 'St Simeon'; it does not even really resemble the 'Vision of the world' which he would describe a few years later in 'Locksley Hall', in which the speaker does not choose whether to imagine the world's future as positive or negative, but merely finds his imagination responding to his shifting moods (Tennyson 1987, 2:125–7). It is also very different from the way in which Hallam had described the operation of the poetic imagination in his 1831 essay, in which the 'poet of sensation' 'live[s] in a world of images' whether they want to or not, rather than only experiencing such images when they deliberately choose to call them to mind (Hallam 1862, 296). Precisely because the speaker *knows* that his 'phantom fair' is entirely imaginary, the spiritual consolation that she can offer him is strictly limited. Such a form of imagination might win the approval of contemporary mad-doctors, who took for granted that '"imagination" unchecked by the will' could be a cause of insanity; but it is hard to imagine it becoming a genuinely creative force (Colley 1983, 23).

By the early 1840s, however, it was widely believed that the alternatives to such self-control could be dire. In 1841, Thomas Medwin, who had known both Byron and Shelley in Italy, wrote a story for *Bentley's*

Miscellany: entitled 'Sydney: From the Memoranda of a Physician', it told the story of a sensitive young poet's decline into delirium and death following a romantic disappointment. Sydney, who was obviously modelled on Shelley, is presented as a firm believer in the superiority of poetic inspiration to ordinary reason: he maintains that true genius transcends time and space, believes in mesmerism and Swedenborgianism, and dismisses the idea that his apparently inspired dreams might be mere madness, asserting like Blake that 'the many consider as madmen, those whose minds are not cast in the same mould as their own'. Yet the last word is not given to him, but to the nameless 'Physician' within whose narrative his speeches are enclosed, and who reflects ruefully that when it comes to madness, 'the man of genius, the poet, the philosopher, those gifted with the most refined sensibilities, are most liable to its fatal influence' (Medwin 1841, 168–71).

Watching Sydney's decline, the physician keeps an open mind, remarking: 'Who can say that he is mad? Perhaps, dead as he seems to the external world, he understands the mysteries of another'. Yet this noncommittal 'perhaps' is small consolation for the suffering which Sydney endures because of his poetic constitution, which afflicts him with sleepwalking, credulity, academic failure, physical weakness, and ultimately premature death. Exposure to his mind, the physician remarks, 'produces on me the effect of opium-eating', thus aligning his poetic fancies with the visions produced by the over-consumption of opiates: intoxicating, no doubt, but ultimately unhealthy and delusive, and founded in physical malfunction rather than any higher spiritual reality (Medwin 1841, 168–74). By 1841, Medwin had built a reputation as something of an authority on poetic genius; and I find it extremely suggestive that in this story he presented that authority in medical terms, as though implicitly accepting Madden's assertion that only medical doctors were truly qualified to understand the poetic mind. In reality, Medwin had been Shelley's friend: but in his story, the narrator is both Sydney's friend ***and his physician***, and is thus able to speak with authority about the way in which his 'refined sensibilities' contribute to his early death.

Collectively, these works bear witness to the ways in which 'inspired' or 'visionary' imagination was systematically pathologised during the years leading up to the passage of the Second Lunacy Act. The widening definitions of insanity that were used at the time, which necessitated the construction of so many new asylums and allowed entire movements such as Tom's rebellion to be written off as 'the Madness that dwells in the hearts of men', also encouraged writers such as Medwin to see any kind of imaginative activity as psychologically hazardous unless it could be contained by the rigorous exercise of the will (Carlyle 1837, 3:345). As Winslow made clear, 'men of great genius and sensibility' had to maintain ceaseless mental vigilance to prevent their sanity being undermined by their 'abnormal condition of the nervous system', as had clearly been the case with both John Clare and John Tom (Winslow 1840, 121). And if they proved incapable of such feats of self-control, then Britain's new asylum infrastructure stood waiting to control them instead.

Conclusion: The Poet's Song

All the events which I have discussed in this chapter—Clare's confinement, Tom's rebellion, Tennyson's disastrous entanglement with Dr. Allen, the rise of the new asylum system, and the publication of Winslow and Barlow's guidebooks on how to stay sane—took place during the nine years between the publication of Tennyson's 1833 poetry collection and his two-volume *Poems* of 1842, years which Armstrong has described as being characterised by 'Tennyson's gradual retreat from the daring of [his] earliest poems' (Armstrong 1996, 43). As I have argued, these were also years which saw the enormous consolidation of the 'medico-psychological' approach to the management of insanity in Britain—a consolidation whose most obvious physical manifestations were the gigantic county asylums which, by the 1840s, were being built all over England, and whose most obvious cultural manifestation was the enormous expansion in the kinds of beliefs and behaviours that could be classed as constituting insanity. It is my contention that these two facts are not unrelated, and that if Tennyson was retreating during these years, then the new asylum system may have been one of the things that he was running away from.

As I have discussed, this was a period in which Tennyson's mental health was particularly poor, first due to his depression over Hallam's death, and then due to his anxiety over the failure of Allen's business schemes. It may be for this reason that his poems of 1837–1838 seem so keen to reconcile poetic imagination with the 'secrets of the brain', in order to ensure that writing poetry did not further weaken his grip on sanity (Tennyson 1987, 2:57). If so, then they would, like Medwin's story of 1841, constitute further proof of the extent to which the medicalised accounts of poetic genius offered by doctors such as Browne, Conolly, and Winslow had come to be regarded as authoritative—not least because of the greatly enhanced levels of authority which such medical men had gained by becoming the keepers of the giant new asylums of the period.

When Tennyson published his two-volume *Poems* in 1842, he chose to end the collection with a short poem entitled 'The Poet's Song'. In it, he described a 'Poet' with a capital P, an Orphic figure whose 'melody loud and sweet' has the power to '[make] the wild-swan pause in her cloud, / And the lark drop down at his feet' (Tennyson 1987, 2:177–8). This figure is clearly a kinsman of the visionary poet whom Tennyson had described in his 1830 poems 'The Poet' and 'The Poet's Mind', and his description of how 'A light wind blew from the gates of the sun' as the poet walks out of town in 'The Poet's Song' recalls his description of the mind of a poet as 'Bright as light, and clear as wind' in 'The Poet's Mind' (Tennyson 1987, 1:246, 2:178). But the claims made for the 'Poet' of 1842 are very much smaller than in 1830. In 1830, Tennyson had imagined the poet as one who 'saw thro' life and death', and whose poetry had the power to start world-shaking revolutions:

> And Freedom reared in that august sunrise
>   Her beautiful bold brow,
> When rites and forms before his [i.e. the poet's] burning eyes
>   Melted like snow.
>
> [...]
>
>   No sword
>   Of wrath her right arm whirled,
> But one poor poet's scroll, and with *his* word
>   She shook the world. (Tennyson 1987, 1:243, 1:245)

This really *was* the way John Tom had spoken, as he tried to gather support for his doomed revolution. ('I am the Saviour of the World!' 'The Lord has chosen me for a body from the foundation of the world, wherein God himself would dwell, to maintain Truth, and execute Justice towards the sons of men!') (Rogers 1962, 90–103, 113, 117–8). But by 1842 it was no longer the way that Tennyson wrote, or attempted to write: and if the 'Poet' in 'The Poet's Song' still 'sings of what the world will be / When the years have died away', the poem gives no indication that his descriptions of the future should be viewed as having any kind of prophetic authority. His song is 'sweet' and 'gay' and clearly pleasant to listen to; but there is no reason to think that its singer possesses the kind of visionary insight that might allow him to predict the future, or the kind of revolutionary power which might allow him to change it.

Tennyson may have celebrated the progressive unravelling of the 'secrets of the brain' by Allen and his contemporaries, but their steady relabelling of imaginative and political radicalism as madness—not to mention the three years he spent living alongside Allen's asylum—does not seem to have served to increase his confidence in the idea of the poet as a visionary prophet of revolution. Such Shelleyan claims of how the world would shake at the words of the truly inspired poet could sound a great deal like the kind of megalomania which had led John Tom to his miserable fate: and it was surely better to sing songs to charm the birds than to end up like poor John Clare.

References

Newspapers and Journals

The Brighton Patriot. 1835–1839. Brighton.
The Manchester Times and Gazette. 1831–1849. Manchester.
The Morning Chronicle. 1769–1865. London.
The Times. 1785–2019. London.
The York Herald. 1791–1889. York.

Primary Sources

Allen, Matthew. 1837. *Essay on the Classification of the Insane*. London: John Taylor.

Blake, William. 1988. In *Complete Poetry and Prose*, ed. David Erdman, Revised ed. New York: Anchor.

Browning, Robert. 1983–2009. *Poetical Works*, ed. Ian Jack, Rowena Fowler, Margaret Smith, Robert Inglesfield, et al., 15 vols. Oxford: Clarendon.

Canterburiensis. 1838. *The Life and Extraordinary Adventures of Sir William Courtenay, Knight of Malta, Alias John Nichols Tom*. Canterbury: James Hunt.

Carlyle, Thomas. 1837. *The French Revolution: A History*. 3 vols. London: James Fraser.

Clare, John. 1984. *Later Poems*, ed. Eric Robinson and David Powell, 2 vols. Oxford: Clarendon.

———. 2004. In *Major Works*, ed. Eric Robinson and David Powell. Oxford: Oxford University Press.

Conolly, John. 1830. *An Inquiry Concerning the Indications of Insanity*. London: John Taylor.

Hallam, Arthur. 1862. In *Remains in Verse and Prose of Arthur Henry Hallam*, ed. Henry Hallam. London: John Murray.
Madden, Richard. 1857. *Phantasmata*. 2 vols. London: T.C. Newby.
Medwin, Thomas. 1841. Sydney: From the Memoranda of a Physician. In *Bentley's Miscellany*, vol. 9, 168–179. London: Richard Bentley.
Ray, Isaac. 1838. *A Treatise on the Medical Jurisprudence of Insanity*. 1st ed. Boston: Little, Brown, and Company.
Tennyson, Alfred. 1982. *Letters*, ed. Cecil Lang and Edgar Shannon, Jr., 3 vols. Oxford: Clarendon.
———. 1987. *The Poems of Tennyson*, ed. Christopher Ricks, 2nd ed., 3 vols. Harlow: Longman.
Winslow, Forbes. 1840. *The Anatomy of Suicide*. London: Henry Renshaw.
———. 1842. *On the Preservation of the Health of Body and Mind*. London: Henry Renshaw.
Wordsworth, William. 1985. In *The Fourteen-Book Prelude*, ed. W.J.B. Owen. Ithaca: Cornell University Press.

Secondary Sources

Armstrong, Isobel. 1996. *Victorian Poetry*. London: Routledge.
Batchelor, John. 2014. *Tennyson: To Strive, To Seek, To Find*. London: Vintage.
Bate, Jonathan. 2004. *John Clare: A Biography*. London: Picador.
Blackmore, Evan. 1986. John Clare's Psychiatric Disorder and Its Influence on His Poetry. *Victorian Poetry* 24 (3): 209–228.
Bloom, Clive. 2013. *Victoria's Madmen: Revolution and Alienation*. Basingstoke: Palgrave Macmillan.
Colley, Ann. 1983. *Tennyson and Madness*. Athens, GA: University of Georgia Press.
Colville, Derek. 1970. *Victorian Poetry and the Romantic Religion*. Albany: State University of New York Press.
Cooter, Roger. 1984. *The Cultural Meaning of Popular Science: Phrenology and the Organisation of Consent in Nineteenth-Century Britain*. Cambridge: Cambridge University Press.
Cronin, Richard. 2002. *Romantic Victorians: English Literature 1824–40*. Houndmills: Palgrave.
Faas, Ekbert. 1988. *Retreat into the Mind: Victorian Poetry and the Rise of Psychiatry*. Princeton: Princeton University Press.
Faubert, Michelle. 2005. Cure, Classification, and John Clare. *Victorian Literature and Culture* 33 (1): 269–291.
Foss, Arthur, and Kerith Trick. 1989. *St Andrew's Hospital, Northampton: The First 150 Years (1838–1988)*. Cambridge: Granta.

Foucault, Michel. 2006. *History of Madness*, trans. Jonathan Murphy, and Jean Khalifa and ed. Jean Khalifa. London: Routledge.

Halmi, Nicholas. 2004. How Coleridge was Wilder than Byron. *Romanticism* 10 (2): 144–157.

Hayes, Bob. 1987. John Clare and Enclosure. *John Clare Society Journal* 6: 10–19.

Janowitz, Anne. 1998. *Lyric and Labour in the Romantic Tradition.* Cambridge: Cambridge University Press.

Jones, Kathleen. 1955. *Lunacy, Law, and Conscience, 1774–1845.* London: Routledge and Kegan Paul.

Mellett, D.J. 1982. *The Prerogative of Asylumdom: Social, Cultural, and Administrative Aspects of the Institutional Treatment of the Insane in Nineteenth-Century Britain.* New York: Garland.

Melling, Joseph, and Bill Forsythe. 2006. *The Politics of Madness: The State, Insanity, and Society in England, 1815–1914.* London: Routledge.

Pearsall, Cornelia. 2008. *Tennyson's Rapture: Transformation in the Victorian Dramatic Monologue.* New York: Oxford University Press.

Porter, Roy. 1987. *Mind Forg'd Manacles: A History of Madness in England from the Restoration to the Regency.* Cambridge, MA: Harvard University Press.

Ray, Isaac. 1853. *A Treatise on the Medical Jurisprudence of Insanity.* 3rd ed. Boston: Little, Brown, and Company.

Reay, Barry. 1990. *The Last Rising of the Agricultural Labourers: Rural Life and Protest in Nineteenth-Century England.* Oxford: Clarendon.

Ricks, Christopher. 1989. *Tennyson.* 2nd ed. Berkeley: University of California Press.

Rogers, Philip. 1962. *Battle in Bossenden Wood: The Strange Story of Sir William Courtenay.* London: Oxford University Press.

Rosen, George. 1968. *Madness in Society.* London: Routledge and Kegan Paul.

Rylance, Rick. 2000. *Victorian Psychology and British Culture, 1850–1880.* Oxford: Oxford University Press.

Scull, Andrew. 1979. *Museums of Madness: The Social Organization of Insanity in Nineteenth-Century England.* London: Allen Lane.

———, ed. 1991. *The Asylum as Utopia: W.A.F. Browne and the Mid-Nineteenth-Century Consolidation of Psychiatry.* London: Tavistock.

———. 2006. *The Insanity of Place/The Place of Insanity: Essays on the History of Psychiatry.* London: Routledge.

Scull, Andrew, Charlotte Mackenzie, and Nicholas Hervey. 1996. *Masters of Bedlam: The Transformation of the Mad-Doctoring Trade.* Princeton: Princeton University Press.

Showalter, Elaine. 1987. *The Female Malady: Women, Madness, and English Culture, 1830–1980.* London: Virago.

Shuttleworth, Sally. 1996. *Charlotte Brontë and Victorian Psychiatry.* Cambridge: Cambridge University Press.

Smith, Leonard. 1999. *Cure, Comfort, and Safe Custody: Public Lunatic Asylums in Early Nineteenth-Century England*. London: Leicester University Press.

Smith, Cathy. 2006. Family, Community, and the Victorian Asylum: A Case Study of the Northampton General Lunatic Asylum and Its Pauper Lunatics. *Family and Community History* 9 (2): 109–124.

Sournia, Jean-Charles. 1990. *A History of Alcoholism*, Trans. Nick Hindley and Gareth Stanton. Oxford: Blackwell.

Stevenson, Catherine Barnes. 1982. The Shade of Homer Exorcises the Ghost of De Quincey: Tennyson's "The Lotos-Eaters". *Browning Institute Studies* 10: 117–141.

Suzuki, Akhito. 2006. *Madness at Home: The Psychiatrist, the Patient, and the Family in England, 1820–60*. Berkeley: University of California Press.

Tennyson, Charles, and Hope Dyson. 1974. *The Tennysons: Background to Genius*. London: Macmillan.

Whitlock, F.A. 1963. *Criminal Responsibility and Mental Illness*. London: Butterworths.

Wood, Jane. 2001. *Passion and Pathology in Victorian Fiction*. Oxford: Oxford University Press.

CHAPTER 5

Seeing Things: Mesmerism, Spiritualism, and Romantic Poetry, 1836–1855

This chapter investigates the rise of mesmerism and spiritualism in Britain, and the attractions that both movements held for authors and admirers of Romantic poetry. Mesmerism, in particular, appeared to offer scientific proof that exceptional individuals might be capable of seemingly supernatural feats of perception and persuasion, just as the Romantics had proclaimed. The relationship between mesmeric trance states and poetic creativity will be discussed in relation to the writings of J.C. Colquhoun, Elizabeth Barrett, Harriet Martineau, and Chauncey Hare Townshend, all of whom associated the powers of mesmerism with those of the imagination. It compares their pro-mesmeric writings to those of Browning, arguing that his hostility towards both mesmerism and spiritualism was strongly connected to his scepticism regarding visionary poetry, and explores the roles played by mesmerism and poetic inspiration in Barrett's and Browning's courtship correspondence. The differing ways in which Barrett and Browning responded to mesmerism and spiritualism, and to the disastrous séance that they both attended in 1855, serve to illustrate the ways in which beliefs about mesmerism and spiritualism were deeply entangled with contemporary controversies over the nature of poetic genius.

© The Author(s) 2019

J. Crawford, *Inspiration and Insanity in British Poetry*,
Palgrave Studies in Literature, Science and Medicine,
https://doi.org/10.1007/978-3-030-21671-9_5

Phenomena: Mesmerism and Spiritualism in Britain, 1836–1855

On 23 July 1855, Robert and Elizabeth Barrett Browning attended a séance held by the famous medium Daniel Dunglas Home. Home was not the first spiritualist medium to become a celebrity, but he was one of the most accomplished: by 1855 his achievements had extended far beyond communication with the dead, and included levitating off the floor, elongating his limbs, moving objects without touching them, and causing entire rooms to shake. Several other mediums of the period also enacted such 'phenomena', as they came to be called, but few were as spectacular as Home; and while many such performers were eventually exposed as frauds, the method by which Home accomplished his feats were never adequately explained. Pope Pius IX was so disturbed by Home's apparently supernatural powers that he had him expelled from Rome on a charge of sorcery (Lamont 2004, 899–902). After attending Home's séance, Robert Browning was convinced that everything he had seen had been the result of trickery. But Elizabeth Barrett Browning was not persuaded: the only plausible explanation, she maintained, was that Home really had been in touch with spiritual forces which transcended the ordinary physical world (Dally 1989, 167–8; Thomas 1982, 157–9).

Originating in America in 1847–1848, the craze for spiritualism had arrived in Britain in 1852, and had rapidly taken the country by storm. However, the way for it had to a great extent been prepared by the earlier controversies over mesmerism, which had generated such extraordinary claims and counter-claims during the 1830s and 1840s. Many of the earliest and most ardent British converts to spiritualism were men and women who were already committed believers in mesmerism: and for people who already accepted the reality of telepathy, clairvoyance, clairaudience, and precognition, accepting the validity of spirit communication often required only a small additional effort of belief (Barrow 1986, 130–2). Indeed, for some it required no additional effort at all: some mesmerists, such as Louis-Alphonse Cahagnet, had been using mesmeric methods to communicate with what they believed to be the spirits of the dead for years before the rise of American spiritualism (Cahagnet 1851). For individuals such as Cahagnet, the spiritualist gospel thus came less as a stunning new revelation than as a welcome confirmation of their existing practices and beliefs.

In Britain, the high-water mark of enthusiasm for mesmerism was reached in the 1840s. During these years its practitioners spliced it together with spiritualism, Swedenborgianism, and phrenology, creating an intoxicating variety of exciting but intellectually rickety pseudo-sciences; but as mesmeric miracle workers came and went and the world remained unpersuaded, the enthusiasm of many of its devotees waned. The publication of John Elliotson's mesmeric and phrenological journal *The Zoist*, which first appeared in 1843 and was discontinued in 1856, serves as a useful marker of the movement's rise and fall (Ruth 1999, 300; Hughes 2015, 135–6). As the appeal of 'pure' mesmerism declined many of the more mystically inclined mesmerists transferred their loyalties to the spiritualist movement, whose star was on the ascendant following the publication of Andrew Jackson-Davis' spiritualist-mesmerist 'bible', *The Principles of Nature*, in 1847, and the widely publicised séances held by the Fox sisters in 1848. Spiritualism, in turn, flourished in Britain throughout the 1850s and 1860s, and retained a considerable popular following well into the twentieth century.

By using these mesmeric and spiritualist 'sciences', the clairvoyants, mediums, and somnambulists of the period claimed to be able to perceive all manner of extraordinary things: spirits, angels, the souls of the dead, the past, the future, scenes that were occurring many miles distant, the contents of other people's minds, and so on. Many people who attended their performances also reported witnessing strange sights: floating objects, spinning tables, spirit manifestations, levitating men. Some were so startled by what they saw that they went in sceptics and came out believers. One of the Masters in Lunacy, Charles Lockhart Robertson, confidently asserted that Home's 'phenomena' could have no existence except in the over-stimulated imaginations of his audience: but after attending one of Home's performances himself in 1857, he reversed his view, concluding that the scenes which he had witnessed could not possibly have a purely psychological explanation (Lamont 2004, 900). The mesmeric pioneer John Elliotson was said to have been so amazed by what he witnessed in Home's séances that he abandoned his earlier belief in materialism and converted to Christianity (Barrow 1986, 95).

Such visions and wonders had, of course, been the stock-in-trade of miracle workers throughout history. What was new about mesmerism and spiritualism was their enthusiastic engagement with the culture of nineteenth-century science, which their adherents hoped would offer them unprecedented opportunities to demonstrate the objective validity of their beliefs. When the Virgin Mary appeared to Bernadette Soubirous in 1858, and proceeded to work a variety of healing miracles at Lourdes, no Catholic authority suggested that her miracles needed to be reproduced under laboratory conditions before they could be accepted as valid: the Mother of God was under no obligation to conform to the rigours of the scientific method. But many mesmerists and spiritualists not only accepted but invited such scientific scrutiny, hoping to thereby demonstrate the truth of their claims. Cahagnet even offered to submit himself for medical treatment as a lunatic if the proofs he offered for spiritualist mesmerism could be scientifically disproved (Cahagnet 1851, 119). This eagerness to seek the support of the physical sciences arose largely out of a desire to rebut the most obvious objection to their assertions, namely, that the extraordinary 'phenomena' which they claimed to have witnessed might have been delusions or hallucinations. Someone who claims to see spirits and auras may just be crazy: but their experiences become much harder to explain in purely psychological terms if they also appear to be demonstrably capable of seeing through walls, or of levitating objects on demand.

Valuable though these demonstrations were in winning converts for the mesmeric and spiritualist movements, in a larger sense they represented a compromise with the prevailing scientific world-view of the era. For mesmerists and spiritualists alike, the mere testimony of visionaries evidently no longer provided sufficient grounds for accepting the truth of their visions: instead, it had to be supported by empirically observable external phenomena, which provided apparent proofs that the spiritual forces they perceived had some real existence outside their own minds. This reliance upon empirical verification set them apart from the followers of earlier self-proclaimed prophets such as Swedenborg, John Wroe, or Johanna Southcott (not to mention visionary poets such as Blake, Wordsworth, and Shelley), all of whom maintained that their respective messages should be able to persuade those who heard them on their own merits, rather than because their speakers were also able to levitate tables or read letters without removing them from their envelopes. In *Milton* (1804), Blake had written contemptuously of those who would not believe the words of prophets unless they were supported by obviously supernatural occurrences:

> No Faith is in all the Earth: the Book of God is trodden under Foot:
> He sent his two Servants Whitefield & Westley [sic]; were they Prophets
> Or were they Idiots or Madmen? shew us Miracles!
>
> Can you have greater Miracles than these? Men who devote
> Their lifes whole comfort to intire scorn & injury & death
> Awake thou sleeper on the Rock of Eternity Albion awake. (Blake 1988, 118)

For Blake, 'shew us Miracles!' was, in all ages, the cry of the spiritually obtuse: in the gospels, it was the unbelieving scribes and Pharisees who said to Jesus 'we would see a sign from thee' (Matthew 12:38). From such a perspective, by offering to perform miracles on demand, the mesmerists and spiritualists might win individual battles against unbelief, but they had already lost the war. Mesmerism, especially, served more to reinforce the contemporary tendency to medicalise all odd and abnormal forms of human behaviour than to challenge it. Indeed, as I have discussed, mesmerism appealed to writers such as Colquhoun precisely *because* it promised to bring the chaotic fields of miracle and prophecy into the domain of testable medical science.

However, mesmerists and spiritualists alike swiftly discovered that their attempts to offer empirical evidence in support of their beliefs were not as persuasive as they had hoped, as the very phenomena which believers found so convincing were dismissed by sceptics as the results of delusion, imagination, or fraud. When the pro-mesmeric writer Catherine Crowe published *The Nightside of Nature* in 1848, containing what she asserted to be several hundred verified narratives of apparently supernatural events, she insisted that they could not *all* be cases of delusion, hallucination, or deception, but her critics were not persuaded by the immense accumulation of data with which she presented them (Crowe 1848, 1:20–1). Just as the increasingly expansive definitions of insanity used in mid-century Britain allowed entire religious or political movements to be written off as the result of mass delusion, so the testimony of believers in mesmerism and spiritualism, be they ever so numerous, was always vulnerable to being dismissed as simply the result of group hysteria deliberately manipulated by devious charlatans—especially because mesmerism and spiritualism, like poetry, were understood to threaten the mental health of their practitioners, because all three practices involved the dangerous over-stimulation of the imagination. When the doctor and sceptic James Edwards explained in 1871 why he did not believe the claims of spiritualism, despite having attended many séances (including those of Home), he concluded that '[spiritualist phenomena] have their origin in unconscious action or self delusion, unless they are the result of imposture'. He furthermore asserted 'that such *séances* are extremely injurious to delicate persons of sensitive organisation, and that they tend to unhinge the mind', gloomily noting that several of the spiritualists he had met over the years had since declined into insanity (London Dialectical Society 1871, 80–1, 76).

Over the last 35 years, the interlinked histories of mesmerism and spiritualism in Britain have been ably researched by a number of scholars, including Janet Oppenheim, Logie Barrow, Alex Owen, Alison Winter, Marlene Tromp, and William Hughes. My aim in this chapter is to draw upon their works in order to explore the ways in which these histories overlapped with that of post-Romantic poetry in Britain, thus filling in some of the historical background to the 1855 encounter between Robert and Elizabeth Browning and Daniel Dunglas Home. In particular, I shall discuss the works of J.C. Colquhoun and Chauncey Hare Townshend, who, though little-known today, were among the foremost advocates of mesmerism in Britain during the 1830s and 1840s. They were also both admirers of Romantic poetry, which they drew upon in order to help them articulate the nature of the altered states of consciousness unlocked by mesmeric somnambulism: and, as a result, their works provide particularly clear illustrations of the relationship between mesmerism and the cultural legacies of British Romanticism.

'Come, Bow Your Souls Before the Bard!': The Mesmeric and Poetic Ambitions of Chauncey Hare Townshend

Mesmerism held a special attraction for the authors of the early Victorian period, and Robert Browning was unusual in remaining sceptical of the claims of mesmeric science. Charles Dickens became an active practitioner of mesmerism, and Harriet Martineau wrote strongly in its favour after being restored to health by mesmeric treatment in 1844. George Eliot, Wilkie Collins, Edward Bulwer-Lytton, and Charlotte Brontë were all fascinated by mesmerism, and depicted it in their respective fictions, while Tennyson experimented with mesmerising his wife, Emily, as a way of providing her with relief from pain (Connor 2010, 2, 7–10; Chapman 2003, 49–50; Colley 1983, 42).

The abilities of mesmerists had obvious parallels with those of authors. Both aimed to captivate the attention of their subjects, direct their thoughts, evoke new emotions within them, and allow them to see, in their mind's eye, people and places which were not actually present. But the larger claims which were made for mesmerism—that it could heal the minds of its subjects, or allow them to look across time and space, or let them perceive hidden, spiritual realities—were also reminiscent of the claims made for the power of poets and poetry by writers such as

Wordsworth and Shelley, whose newfound popularity in the 1830s coincided with the rising fortunes of mesmerism in Britain. The Romantics themselves had been fascinated by mesmerism: when Edward Irving visited Coleridge in 1824, he reported that the poet was full of 'poetry and animal magnetism', while Shelley's interest in mesmeric healing is clear from his 1822 poem 'The Magnetic Lady to her Patient', although his attempts to have Mary Shelley magnetise him into greater literary creativity had to be terminated after he attempted to jump out of a window while mesmerised (Sanders 1942, 150; Handley 2012, 317; Ruston 2013, 64–5). Mesmerism, like Romantic poetry, held out the promise of allowing its practitioners to become seers or prophets, opening the eyes of the unbelieving world to truer and higher things. Blake had written of 'the Bard! / Who past, present, and future sees', and in the early Victorian period many mesmerists claimed to have gained access to a very similar set of abilities, even if they sought to explain them in more medicalised terms (Blake 1988, 18).

It is thus surely no accident that one of the earliest and most enthusiastic advocates of mesmerism in Britain, Chauncey Hare Townshend, was also a would-be Romantic poet, who as a young man sought out meetings with Southey, Wordsworth, Coleridge, and Clare (Kaplan 1975, 58). He was sending poems to Southey as early as 1815, when he was just 17 years old—a time in his life when, as he would later recall, his soul was full of 'poet-love' (Townshend 1851, 26–8). As a student at Cambridge he won the 1817 Chancellor's prize for his poem 'Jerusalem', and his first collection of *Poems*—which was dedicated to Southey—was published in 1821, the same year in which he met John Clare (Matthew and Harrison 2004, 55:152; Bate 2004, 183–5). Like Tennyson nine years later, Townshend had an extremely exalted opinion of poetry, and in 'The Poet's Passion' he asserted the superiority of poetry to all other callings, claiming that the 'persuasive arts' of the poet were capable of accomplishing greater conquests than any soldier, and of building greater empires than any king. 'Warrior, chief, miser, lover—all,' he demanded, 'Come, bow your souls before the bard!' (Townshend 1821, 173). In his 'Ode on the Diversity of the Poetic Character', he asserted that the very greatest of poets were capable of even more superhuman feats:

> For him the shadowy curtain is withdrawn
> That veils the unknown world from mortal sight,
> But there, for him, unfading glories dawn,
> While angels beckon to the realms of light. (Townshend 1821, 88)

Despite (or perhaps because of) their immense superiority to other men, however, Townshend viewed true poets as being extremely vulnerable to misunderstanding and neglect. In 'The Poet's Grave', a poem which was probably inspired by the death of John Keats, Townshend imagined himself as mourning 'on the Bard's untimely grave', lamenting the fact that 'Fate's sordid smile / Beams on the heartless and the vile / While merit weeps alone'. He is finally comforted by a vision in which the poet's spirit appears in glory, assuring him that whatever neglect true poets may experience on earth, in heaven 'Justice rules, and makes the balance even!' (Townshend 1821, 52, 54, 56).

In both his 'Ode on the Diversity of the Poetic Character' and 'The Lonely Heart', Townshend wrote that poets of a sensitive and solitary disposition, among whom he clearly numbered himself, were especially ill-fitted to the world:

> More noble, far, the silent woe
> Of him, who finds few joys below,
> Who ne'er his feelings can express,
> Who is not form'd for happiness;
> Not only poet, but endued
> Above the tuneful multitude,
> With the poetic-temper'd frame
>   Respondent to the soul it shrines,
> Which, with a strength, no force can tame,
>   Too feeling tenderness combines;
> Thro' every shrinking pore alive,
> In every fibre sensitive;
> Fit medium for th' electric fire,
> Which, ever-restless, will aspire,
> Beyond mortality's allowed desire. (Townshend 1821, 83)

Anticipating the arguments which would be put forward by Horne, Hallam, Mill, and Madden in the early 1830s, Townshend asserted that such poets are born, not made. Even more than other kinds of poets, they are 'endued' with 'poetic-temper'd frame[s]' and 'fibre[s]' of extreme sensitivity: as a result, their bodies are 'fit medium[s] for th' electric fire', by which Townshend presumably meant that the galvanic electrical fluids which were thought to animate all life flow through them with particular force, driving them towards superhuman goals that extend 'Beyond mortality's allowed desire'. Their powerful imaginations allow them to envision 'That bright ideal excellence / Which on the inward eye hath shined', and because of this they are always doomed to misery, as nothing that they can perceive with 'the outward sense' can ever live up to this ideal (Townshend 1821, 84).

Townshend was probably thinking of *Alastor* when he wrote these lines, and his description of this poetic type is more or less a plot summary of Shelley's poem. But his description of a very similar-sounding case in 'The Lonely Heart' is clearly meant to be autobiographical:

> There is a joy in loneliness,
> Which lonely minds alone can know,
> Such as to none can e'er express
> The secrets of their joy or woe;
>
> Souls, wild, and various as the lyre,
> That ne'er to mortal touch will yield;
> Mysterious as the tomb's deep fire,
> Never to mortal eye reveal'd;
>
> Who feel within them deathless powers,
> That pant, and struggle to be free;
> That would outstrip Time's lazy hours,
> And launch upon Eternity.
>
> Ah, little deem the blind, dull crowd,
> When gazing on a tranquil brow,
> What thoughts and feelings unavow'd,
> What fiery passions lurk below! (Townshend 1821, 176)

Every cliché of Byronic or Shelleyan poetic genius is here: the great, 'Mysterious' soul with its 'deathless powers', which can only be understood by other 'lonely minds' like itself, and is thus doomed to be misunderstood by the 'blind, dull crowd'. The reason that such 'lonely minds' are lonely, one gathers, is not because they are socially awkward but because their 'deathless powers' and 'fiery passions' set them apart from the rest of humanity, leaving them longing to 'launch upon Eternity'. Townshend goes on to describe how, while they are saying 'custom's trivial phrase', such individuals may in fact be 'Feed[ing] sweetly on some chosen theme' or 'Weav[ing] the light tissues of a dream', and he goes on to describe in ecstatic terms the 'rapture', 'strange disordered transport', and 'thousand visions' which they may inwardly experience, all entirely unguessed at by those who see only their 'outward frame'. Like the 'born poets' described by Hallam, these 'lonely minds' can appreciate colour, nature, and music on a much deeper level than most people, gaining from them 'bliss, beyond the vulgar joy' (Hallam 1862, 295). Their greatest pleasures, however, are purely inward: and when they shut their eyes, 'the bright soul opes her eyes within', showing them 'scenes of more than mortal hue'.

Being thus enraptured by these solipsistic pleasures, and bearing a spirit which 'weeps at what itself conceived', such people will inevitably disdain all earthly things:

> These joys are his: and can he turn
>   One thought upon the world below?
> Can he do aught but proudly spurn
>   All it can promise, or bestow? (Townshend 1821, 176–80)

Like Browning's Sordello, or the speaker of *Pauline*, Townshend's 'Lonely Heart' clearly thinks that his dreams and feelings, in and of themselves, are of such exalted quality that they mark him out as an immensely superior individual, even though he himself acknowledges that he has no way of comparing them with anyone else's. These dreams and feelings do not need to take any outward form, such as written poetry, in order to become valuable; indeed, Townshend appears to prefer the idea of them remaining purely internal, completely hidden from 'the blind, dull crowd'. He seems entirely unaware that the type he describes might not be an attractive one, or that their loneliness may be partly due to the fact that people can tell when they are being fobbed off with 'custom's trivial phrase' by someone who isn't actually paying attention, and do not much enjoy talking to individuals who are obviously daydreaming rather than listening.

As a young man, then, Townshend was an early and enthusiastic adopter of the idea of the poet as visionary genius, whose 'poetic-temper'd frame' gave him or her access to realms of experience entirely inaccessible to lesser mortals. Such individuals *might* attain recognition and fame, but as 'The Poet's Grave' made clear, they probably wouldn't: most people would never understand their true merit, and they themselves would be constitutionally incapable of 'turn[ing] / One thought upon the world below' to make themselves better understood. He was thus probably not entirely surprised when his 1821 *Poems* was largely ignored by readers and critics. His family's wealth was sufficient for him to be able to devote himself to literary pursuits, and throughout the 1820s and 1830s he continued to publish short works in poetry and prose.

Townshend's writings suggest, however, that by the later 1820s he had recognised that not every person who claimed to be a wild and lonely soul was necessarily a great poet in the making. His satirical four-book poem *The Reigning Vice* (1827) criticised the selfishness of individuals who presented themselves as solitary and superior beings simply as an excuse for evading their real moral duties; it also attacked the prevailing fashion for gloomy Byronic poetry, arguing that it encouraged egotistical self-obsession among poets who competed to describe 'How *I* detest the world, and from it flee, / Because the world detests, and flies from *me!*' The poem still has high praise for 'the few, who scorn what earth can give'; but Townsend clearly viewed such genuinely world-denying individuals as being few and far between, and certainly much smaller in number than the group who *claimed* such status simply on the basis that they were moody and gloomy and alone (Townshend 1827, 11, 62–3, 115).

In a somewhat similar vein, Townshend's long 'Essay on the Theory and Writing of Wordsworth' (1829), which was published in four monthly parts in *Blackwood's Magazine*, asserted that one of the major problems with Wordsworth's philosophy of poetry was that it was insufficiently elitist. Over the course of the essay, Townshend variously criticised Wordsworth for claiming that elevated thoughts and sensitivity to natural beauty could be found among the rural poor, for depicting a common peddler as speaking like a philosopher, for asserting that the ordinary language of everyday life could ever have the same power as the language of poetry, and for writing poems on such unworthy subjects as drunken waggoners, idiot boys, and impoverished old women. True poetry, argued Townshend, was written by 'inspiration and divine madness', and reflected the 'more lively sensibility of the poet', who 'feels small things more acutely than men of dull and sluggish imagination feel great ones'. As a result, 'the very shadows of [the poet's] mind are stronger than the reality of others'. Townshend even argued that a poem written about a battle by a poet who has only heard about it second-hand is 'actually more true' than the eye-witness testimony of a soldier who fought in it, 'because it is more graphical, and produces on the mind a greater sense of reality'. In Townshend's view, one of Wordsworth's great errors was in imagining that his poetical vision of life and nature could be shared by ordinary people, rather than understanding it as a sacred gift which would always set him apart from other men (Townshend 1829, 453–62, 604–9, 780, 786).

Collectively, the works published by Townshend during the 1820s give some strong hints as to why mesmerism appealed to him so powerfully when he first encountered it in 1836. As he recounted in *Facts in Mesmerism* (1840), until the winter of that year he had been 'an unbeliever in the very existence' of mesmeric trance states, and indeed he had mocked such 'magnetic wonders' in his 1829 essay on Wordsworth (Townshend 1840, 56; Townshend 1829, 604). During 'a residence in Antwerp', however, he witnessed a 'Mr K—' induce mesmeric trance states in a series of individuals, in which they not only became obedient to his commands, but also appeared to display clairvoyant abilities. This demonstration left Townshend completely persuaded of the reality of mesmerism, and 'having, through the kindness of Mr K—, received such instructions as put it in my own power to practise the mesmeric art', he soon became an enthusiastic mesmerist himself (Townshend 1840, 56–80). His initial efforts to persuade his English friends of the reality of mesmerism met with very mixed success, but he persevered with his demonstrations; and by the time the controversy over Elliotson's use of mesmerism at University College Hospital broke out in 1838, Townshend had become one of the most high-profile advocates of mesmerism in Britain (Townshend 1840, 88–116).

For Townshend, the practice of mesmerism provided a shortcut to the kind of feats of illumination which he had been dreaming of ever since his days as a poetry-loving adolescent two decades before. The mesmerist, like the visionary poet, could lift 'the shadowy curtain' which 'veils the unknown world'; but whereas poets had to contend with the hostility and indifference of a world which endlessly misunderstood them, the talents of the mesmerist could be objectively demonstrated. The quality of the 'electric fire' which burned within them could be proven by their feats of mesmerism—feats which Townshend believed to offer such clear proof of their abilities that they could not be doubted by any honest witness. Under the influence of Townshend's mesmeric powers, the people mesmerised by him demonstrated 'a new sense of spiritual things', became entirely truthful, and exhibited an improved moral character. The proud became humble, an 'unsentimental' infidel was filled with 'inspiration' and professed to believe in God, and a woman otherwise 'utterly insensible to poetry' was, while mesmerised, so moved by hearing Townshend read Wordsworth that she burst into tears (Townshend 1840, 165–71). Everything, in other words, that Townshend had initially hoped to accomplish by writing poetry could, with the aid of mesmerism, be achieved almost literally with a wave of his hand.

The strong connection which, for Townshend, existed between poetry and mesmerism is clearly exhibited by the frequency with which he quoted Wordsworth (and Milton) in order to explain the nature of the mesmeric trance states which he induced in other people. The mesmeric state, he wrote, 'half-realises to us the inspired words of the poet' in Wordsworth's 'Intimations of Immortality'; it brings us closer to Wordsworth's 'fountain-light of all our day', permits access to the 'serene and blessëd mood' described by Wordsworth in 'Tintern Abbey', and proves that 'The external world is fitted to the mind', just as Wordsworth had argued it was. Mesmeric force was none other than the 'active principle' ascribed to 'every form of being' in Wordsworth's *Excursion*, and its influence helps us to 'see by the glad light / And breathe the sweet air of futurity', as recommended in the same poem: but, first, in order to get the best results, the patient to be mesmerised should cultivate that state of mind 'which Wordsworth has characterised under the title of "a wise passiveness"' (Townshend 1840, 189–90, 357, 409, 429, 432, 469, 524). Mesmerism, in other words, ultimately achieved results identical to those of Wordsworth's poetic philosophy, but with the important distinction that it allowed the entire process of spiritual development described in Wordsworth's poetry to be short-circuited. Rather than gradually developing a spiritual understanding of the universe over the course of many years of suffering, the patient could be instantaneously shifted into that happy state simply by having the good fortune to be mesmerised by Townshend. In one particularly bizarre episode, Townshend even described an occasion in which he mesmerised a woman without even looking up from a book of Wordsworth's poems which he was reading at the same time! (Townshend 1840, 456–7).

In theory, the mesmeric powers described by Townshend were supposed to be accessible by almost anyone, making them very different from the kind of poetic talents which, in the 1820s, he had described as being available only to a tiny elite. In practice, however, Townshend's mesmeric world was strongly hierarchical, divided into the mesmerisers, who were usually older, high-status, and/or male, and the mesmerised, who were usually younger, lower-status, and/or female. (A similar sense of natural hierarchy led Harriet Martineau, who in 1844 was successfully mesmerised by her maidservant, to seek out a mesmeric healer of her own class—a clergyman's widow—at the first opportunity [Martineau 1845, 11].) Among his whole exhaustive catalogue of mesmeric experiments, Townshend does not at any point mention allowing anyone else to mesmerise *him*, and, tellingly, Charles Dickens—whom Townshend met at Elliotson's house in 1840—refused to permit Townshend to mesmerise him for fear that his mind might be damaged by the experience, even though Dickens made liberal use of his own mesmeric abilities on other people (Connor 2010, 10; Kaplan 1975, 65). Likewise, when Townshend introduced Bulwer-Lytton to mesmerism, the latter seamlessly integrated his belief in his own mesmeric powers into his sense of himself as a spiritually superior 'Poetical Character', set apart from other men by his 'visionary' artistic gifts and aristocratic heritage (Mitchell 2003, 132–3, 137, 144). Townshend, Dickens, and Bulwer-Lytton all seem to have viewed themselves as people whose role in the world was to direct mesmeric power upon others, rather than to have it directed towards themselves—a role very much in keeping with Townshend's adolescent self-image as an individual whose 'poetic-temper'd frame' blessed him with mysterious powers that 'the blind, dull crowd' could never equal or understand.

In his 1845 critique of mesmerism, *Mesmerism True, Mesmerism False*, John Forbes singled out Townshend as an example of the way in which mesmerism appealed to the vanity of its practitioners. Quoting Townshend's description of how 'the mental enthusiast' will 'riot in the display of energies so long concealed, so wondrously developed', and of how 'the pride of exerting an influence over others awakens in his breast', so that 'the more incredulous he has hitherto been, the greater will be his rebound to credulity', Forbes commented: 'a perusal of the Reverend author's entire work forces the inference that he is himself the great sublime he draws' (Forbes 1845, 50). Townshend himself had recognised that one of the reasons why mesmerism had been so slow to win general acceptance was because of the strength of its appeal to 'visionaries', and 'person[s] of imagination' who were 'tired of rough reality', which made it easy for its critics to dismiss the entire practice as a mass of overheated fantasies (Townshend 1840, 19, 35).

In this respect, the controversy over mesmerism strongly paralleled the controversy over Romantic poetry which had unfolded a few years earlier. As I have mentioned, in 1833, Mill had written to Carlyle that poetic truths would seem to be 'nothing but dreaming or madness' to most people; similarly, in 1840, Townshend wrote that even when stated with the 'utmost caution', an account of mesmerism would seem like 'madness' to most non-believers (Townshend 1840, 45; Mill 1981, 12:163). In 1834, Taylor had critiqued 'the PHANTASTIC school' of Shelleyan poetry for making grand claims about reality based on nothing more than fantasy and wishful thinking; in 1845, Forbes made much the same objection to mesmerism. (The same arguments would come up yet again during the 1850s controversies over spiritualism.) For a poet such as Townshend, mesmerism could serve as a supplement or a substitute for the kind of visionary bardic power which he had longed for as an undergraduate, and the fact that the mesmerists interpreted this power in thoroughly medicalised terms was unlikely to have seemed off-putting for someone who had been attributing his poetic abilities to the sensitive 'fibres' of his 'poetic-temper'd frame' as early as 1821. To sceptics such as Browning, however, the new mesmeric 'sciences' appeared in a much more doubtful light: and his doubts, as I shall discuss, stemmed at least partly from his increasing wariness of the dangers of unrestrained imagination.

The Mesmeric Imagination

To a very great extent, the debate over mesmerism was also a debate over the power and validity of the imagination. Individual mesmerists might plausibly be charlatans, but as the popularity of mesmerism grew, it became increasingly obvious that such an explanation could hardly account for the mesmeric movement as a whole. Even if Mesmer himself had been a fraud, by the 1840s it beggared belief that *every* advocate of mesmerism—including doctors such as Elliotson, lawyers such as Basil Montagu and J.C. Colquhoun, and clergymen such as Townshend and George Sandby—could all be part of a conspiracy to delude the public. Sceptics such as John Forbes (and Robert Browning) thus resorted instead to a psychological explanation: the mesmerists might sincerely believe themselves to have witnessed or experienced all manner of extraordinary things, but these apparent miracles were due to the power of the imagination rather than the reality of mesmerism (Hughes 2015, 82–4; Kurshan 2006, 26–7).

As Colquhoun, Townshend, Sandby, Martineau, and De Quincey all complained, 'imagination' rapidly became a catch-all explanation for everything the opponents of mesmerism found difficult to account for (Townshend 1854, 6–7, 46; Martineau 1845, 41; Ruston 2013, 95). In such critiques, 'imagination' (or sometimes 'hysteria' or 'suggestion') was held to be capable not only of inducing trance states and visionary experiences, but also of suppressing pain and curing mental and physical diseases.

As a result, it risked becoming every bit as mysterious as—or even effectively identical to—the phenomenon which it supposedly explained. In *Isis Revelata*, Colquhoun objected that people who attributed the powers of mesmerism to 'imagination' never properly defined what this seemingly omnipotent faculty of imagination actually *was*, and cited Chardel's complaint that just as superstitious people used to attribute everything they didn't understand to the supernatural, so contemporary scientists ascribed all such mysteries to 'imagination', effectively attributing 'supernatural powers' to it in the process (Colquhoun 1836, 1:122, 1:284–6, 1:290–1). Similarly, in his 1844 book *Mesmerism and Its Opponents*, Sandby noted that if all the cures accomplished by mesmerists had in fact been due to imagination or 'hysteria', then hysteria evidently possessed such great curative potential as to constitute an impressive medical resource in its own right. Sandby himself did not find such explanations credible: but he wearily concluded that if people preferred to call the phenomenon under discussion 'imagination' rather than 'mesmerism' then he didn't mind them doing so, as long as this made them more willing to accept that it actually existed (Sandby 1844, 122–32).

If mesmerism consisted of nothing but a harnessing of the power of the imagination, then the imagination appeared to be more powerful than anyone (except, perhaps, a few Romantic poets) had ever guessed. Colquhoun did not accept that mesmeric phenomena could be caused entirely by the imagination of the people affected, objecting that it was quite impossible that 'exciting the imagination' could make someone suddenly fall asleep. However, he acknowledged that a powerful imagination constituted an important resource for the would-be mesmerist, and he wrote at some length of what he called 'this strange power of the imagination', by which a man who possessed 'the power of advanced imagination' could control the thoughts of others. This could take trivial forms, such as a fairground magician imagining one card so strongly that a member of the audience was forced to name that card rather than any other, or more dramatic forms, such as a faith healer compelling the imagination of their patient so powerfully that they recovered from an illness which would otherwise have proven incurable. In its grandest manifestations, this power could even change the shape of history, as when a conqueror secured such powerful imaginative dominance over his followers that he was able to drive them on to overrun empires in his name (Colquhoun 1836, 1:141–53, 1:165–9).

Although a staunch believer in mesmerism, Colquhoun had little time for its props and rituals, which he found overly redolent of the kind of old-fashioned ceremonial magic that he viewed mesmerism as having explained and replaced. In *Isis Revelata* he criticised Mesmer's use of 'mystical' theories, his use of magnets and ceremonies, and his 'affectation of mystery', and noted that while Mesmer's followers in Paris had persisted in using 'magnetised' objects in their treatments, the mesmerists in Lyons had been able to achieve similar results using nothing but the power of faith and will (Colquhoun 1836, 1:213–34, 1:260–2). For Colquhoun, it was this exercise of willpower, by which the stronger mind imposed its imagination upon the weaker, which was the very essence of mesmerism. It was for this reason that his later work, *The Fallacy of Phreno-Magnetism Detected and Exposed* (1843), poured scorn upon the idea that a mesmerist could ever 'prove' the reality of phrenology by channelling mesmeric energy into different parts of a patient's brain. Such a patient, he argued, would simply say or do whatever the mesmerist desired, passively sensing and replicating whatever ideas were being transmitted to them from the overmastering imagination of the mesmerist. As a result, mesmerists who were also phrenologists found their patients acting in accordance with phrenological theory, mesmerists who believed themselves to be surrounded by demons found their patients acting as though possessed, and so on (Colquhoun 1843, 6–13). The capacity of the mesmeric state to reveal the truth about the outside world was thus strictly limited by its tendency to reveal the truth of the mesmerist's imagination and desire, instead.

In this, as in so much else, the mesmerist's dilemma mirrored that of the would-be poet. For Colquhoun, the mesmerist, like the visionary poet, was an individual whose superior mental abilities allowed him to sway those around him through his mastery of 'the power of advanced imagination'. The fact that Colquhoun described Wordsworth as having 'almost prophetically, although unintentionally' described the nature of mesmeric trance strongly suggests that, for him, such poets could be placed alongside mystics, faith healers, exorcists, and great conquerors as examples of people who had actually been practising mesmerism without knowing it all along (Colquhoun 1836, 1:141–2, 1:165, 1:203–8, 2:102, 2:107–8). Yet precisely because of the power of the poet's or mesmerist's imagination, the insights gleaned through the exercise of their abilities were always suspect: what they took to be a revelation of the true nature of reality could easily turn out to be nothing more than their own fantasies, mirrored back to them by the passive psyches of their enthralled audiences. It was on these grounds that Colquhoun rejected the practice of phreno-magnetism, and Townshend, in 1854, attacked the then-popular spiritualist practices of table-turning and spirit-rapping, asserting that these apparent spirits—which only ever seemed to tell people exactly what they wanted to hear—were in fact merely mesmeric phenomena unconsciously generated by spiritualist mediums, who were often 'persons of unsound health and unstrung nerves' (Townshend 1854, 185–209). The medium, the exorcist, and the phreno-magnetist thus function, for Colquhoun and Townshend, as the mesmeric equivalents of Browning's Aprile or Tennyson's St Simeon—visionaries who have lost control over their own imaginations, and who have thus come to believe in the real existence of spirits or demons or phrenological organs when, in fact, these things are merely projections of their own minds masquerading as external realities.

Viewed from a less sympathetic perspective, of course, much the same could be said about mesmerism in general. As Kaplan notes, critics of mesmerism argued that 'mesmerism was like poetry, an example of the inventive powers of the human imagination' (Kaplan 1975, 25). Colquhoun and Townshend both believed that some forms of mesmeric phenomena, including clairvoyance, precognition, and mesmeric healing, were completely real and valid; but for sceptics such as Forbes, these were simply further examples of imagination run amok. Forbes' final judgement on the matter was that the mesmeric trance state was nothing more than an artificially induced state of hysteria; as for the supposed feats of clairvoyance carried out by somnambulists, he argued that they were so obviously fraudulent that only the pre-existing determination of the mesmerists to believe in their reality prevented them from detecting the deception. In this view, what mesmerism truly demonstrated was the dangerous ability of the imagination to mislead even highly intelligent people into believing absurdities (Forbes 1845, 12–16, 29–32, 46–60). A powerful imagination was a threat to its possessor, because it could dupe them into mistaking fantasies for realities; but a powerful imagination combined with a forceful personality, which was the combination of traits which Colquhoun attributed to truly 'magnetic men' (and which Tennyson had attributed to the poet-prophets of his early poetry), was doubly dangerous, as an individual who possessed both could impress their fantasies not just upon themselves but upon everyone around them, in a form of contagious mass delusion which could have potentially disastrous results (Colquhoun 1836, 1:142).

At the base of this controversy was a clash between two different views of the imagination. For Colquhoun, the imagination was pre-eminently an *active* faculty, reminiscent of Coleridge's definition of 'The Primary IMAGINATION' as a 'living Power' in *Biographia Literaria*; for him, the mesmeric man of 'advanced imagination' was a figure of power, able to imagine new realities into being at will, much like the inspired poet with his mesmeric 'flashing eyes' whom Coleridge had imagined in 'Kubla Khan' (Coleridge 1983, 1:304; Coleridge 1999, 233). For Forbes, on the other hand, imagination was a primarily passive and *reactive* force, of the kind represented by Wordsworth in 'Goody Blake and Harry Gill'. Confronted by certain stimuli, the over-imaginative mind would hysterically manufacture whatever combination of visual hallucinations and psychosomatic physical responses were necessary to conform with whatever it imagined to be happening. This was why, as far as Forbes was concerned, mesmeric medicine had to be rejected *even if it demonstrably worked*: being fundamentally based on lies and delusions, it could only restore health to the body by undermining the health of the mind. Where Colquhoun saw successful mesmerists as being the masters of their own imaginations, Forbes saw them as having been entirely mastered by it.

Given the central role played by the active imagination in Romantic theories of poetry, it is unsurprising that so many poets of the period preferred Colquhoun's position. During the 1830s and 1840s mesmerism offered an alternate intellectual framework within which the imaginative reveries and visionary raptures of poets could be understood: one that was much more sympathetic than most doctors of the period to the idea that vision and imagination could be genuinely positive forces, able to exert a healthy and beneficial influence upon the world, rather than just being milestones on the road to madness. As I have discussed, this connection between faith in mesmerism and faith in poetic imagination can be seen explicitly in the works of Townshend, but, as I shall show, a similar connection can also be seen within the works of Elizabeth Barrett Browning, with consequences that ultimately contributed to her disagreement with Browning over Home's séance in 1855.

'Heavenly Visions': Visionary Perception in the Works of Harriet Martineau, Elizabeth Barrett, and Richard Horne

In 1839, Harriet Martineau was diagnosed with uterine cancer. For most of the next five years she lived as an invalid, confined to her house and cripplingly dependent upon opiates: a frustratingly limited life for a woman who had previously enjoyed a highly mobile and social existence as one of the most prominent writers in Britain. Traditional medicine seemed helpless to relieve her symptoms, and she resigned herself to a life of sickness. In 1844, however, at the suggestion of Basil and Anna Montagu, she underwent a course of mesmeric treatment which seemed to completely restore her health (Martineau 1845, 3–6). Her recovery transformed her into an evangelist for mesmeric medicine; her *Letters on Mesmerism* were widely read after their publication in 1845, and her fame as an author made her one of the movement's most effective advocates (Winter 1998, 224–5).

For Martineau, the effects of mesmerism were literally revelatory. As well as restoring her heath and curing her addiction to opium, it also induced visionary experiences, in which she saw 'phosphoric lights, glorifying every rock and headland'. Her very first experience of mesmerism was of a dream-like state of light and darkness, in which she saw a bust of Isis lit up with spiritual light. (Was she remembering the title of Colquhoun's 1836 book on mesmerism, *Isis Revelata*?) In these trances she experienced an incommunicable state of 'calm translucent intellectuality' so wondrous that, she asserted, if she had been a Catholic she would have believed them to be 'heavenly visions' and mistaken her mesmerist for an angel or a saint. Under its influence, even 'the dirtiest and meanest of the steam tugs' appeared to be 'dressed in heavenly radiance', even though it was 'the last object that any imagination would select as an element of a vision' (Martineau 1845, 7–8, 15–16).

But was it quite true that there was not *any* 'imagination' which would have selected exactly such a mean and humble 'object' as 'an element of a vision'? In her 1838 poem on Wordsworth, Landon had praised him for precisely this ability to make mundane objects seem luminous with 'glory':

> The glory which thy spirit hath
> Is round life's common things,
> And flingeth round our common path,
> As from an angel's wings,
> A light that is not of our sphere,
> Yet lovelier for being here,
> Beneath whose presence springs
> A beauty never mark'd before,
> Yet once known, vanishing no more. (Landon, *Poetical Works*, 1839, 341)

In the *Prelude*, which was finally published in 1850, Wordsworth himself described the 'Convictions' that he had developed in the late 1790s in very similar terms:

> Nature for all conditions wants not power
> To consecrate, if we have eyes to see,
> The outside of her Creatures, and to breathe
> Grandeur upon the very humblest face
> Of human life. (Wordsworth 1985, 253)

Basil Montagu, who encouraged Martineau (and others) to experiment with mesmeric medicine, had been one of Wordsworth's closest friends in the 1790s, and Martineau's mesmeric visions may have reminded Montagu of Wordsworth's visionary insights half a century before. In both cases, the crucial revelation is that even the most unpromising objects—what Landon calls 'life's common things'—can be lit up with 'radiance' or 'Grandeur' if we can only learn to look upon them in the correct way. Like Colquhoun and Townshend, however, Martineau had apparently discovered a short-cut to such transcendent Wordsworthian vision. Rather than passing through a lengthy process of imagination impaired and restored, she had acquired 'eyes to see' simply by being mesmerised by her maidservant.

One of the people to whom Martineau recommended mesmeric treatment was her friend Elizabeth Barrett, who by 1845 had been living with chronic illness for more than two decades. (Another was Charlotte Brontë, who was mesmerised while staying with Martineau in 1850, although 'the result was not absolutely clear' [Shuttleworth 1996, 30].) Like Martineau, Barrett had been managing her condition through the heavy usage of opiates, using a combination of morphine and ether which she referred to as her 'amreeta draught' (Barrett Browning and Browning 1990, 210–11). None of her doctors had been able to successfully treat her complaint, and she was understandably interested in hearing about alternatives. She did not follow Martineau's suggestion to have herself mesmerically treated: she regarded the idea of placing herself entirely under someone else's power with 'horror', and in any case she asserted that 'in cases like mine the remedy has done harm instead of good'. But she wrote in 1844 that she had 'long been a believer' in mesmerism—as well she might have been, given that Elliotson, Britain's leading champion of mesmeric medicine, was also her father's physician—and that she hoped Martineau's case might help to persuade others of its real existence and power (Karlin 1989, 66, 68; Owen 2004, 20; Winter 1998, 234, 238). Such was her belief in mesmerism that when Poe's short story 'The Facts in the Case of M. Valdemar' was first published in Britain, Barrett believed it to be a factual account of a real mesmeric experiment (Sage 1988, 195–6).

In her youth, Barrett had revered Wordsworth, Keats, and Byron, and hoped to attain the same kind of mystic insight which she believed them to have possessed. (She felt more conflicted about Shelley, whose atheism repelled her: in 1841 she complained that 'his poetry glitters and is cold' [Morlier 1990, 143; Blake 1986, 388].) Like Tennyson and Browning, she had filled her earliest works with depictions of poets as inspired visionaries, always striving towards some transcendent understanding of reality. As early as 1826, in her 'Essay on Mind', she described poets as being set apart from other men and women by their ability to sense the 'silent voice' and 'witching spirit' hidden in the landscape, asserting:

> None read, but Poets, Nature's poetry!
> Its characters are trac'd in mystic hand,
> And all may gaze, but few can understand. (Barrett Browning 1914, 42)

In the preface to her 1838 collection *The Seraphim, and Other Poems*, rejecting the idea that poetic imagination might simply be a species of delusive fantasy, she wrote that 'poetry is essentially truthfulness; and the very incoherences of poetic dreaming are but the struggle and the strife to reach the True in the Unknown'. In one of the poems from that collection, 'Earth and Her Praisers', she described a poet as being recognisable from his 'sense of lonely power' and his 'words of fire', concluding that 'in every time or place / Ye may know a poet's face / By the shade or shining' (Barrett Browning 1914, 77, 123). But while both Tennyson and Browning had retreated from this ideal of the poet-mystic by the early 1840s, Barrett, like Townshend, held fast to it, and her strongest defence of such poetry appeared not in her early works but in 'A Vision of Poets', which was published in 1844.

In 'A Vision of Poets', Barrett narrates an allegorical tale in which a poet, kept from sleep by the inner light which burns within him, meets a mysterious lady who requires him to drink from the pools of the world's scorn and bitterness in order to earn a vision of the true poets of the past. (These poets form a predictably Romantic canon: Shelley, Keats, Byron, and Coleridge are all present, but Dryden, Pope, and their poetic successors are pointedly absent, with the roll-call leaping straight from Milton to Burns.) True poets, he learns, must give their lives and hearts for poetry: the poets he sees in his vision have literally cut their own hearts out, giving them over to an angel who plays upon them like a musical instrument, generating divine music in the process. Such poets are scorned and unrewarded by the world, and, like Keats, they usually die young: but they are still 'God's prophets of the beautiful', and the music struck from their hearts makes the earth more glorious in future ages. They are 'poets true / Who died for Beauty as martyrs do / For truth—the ends being scarcely two', and their suffering is the price that they pay for their access to the divine faculty of vision—a price which they gladly pay (Barrett Browning 1914, 215–23). Ultimately the poet in the 'Vision' dies for his art, as all true poets apparently must, and the poem concludes with the lines: 'KNOWLEDGE BY SUFFERING ENTERETH, / AND LIFE IS PERFECTED BY DEATH' (Barrett Browning 1914, 227–9).

In the preface to her 1844 *Poems*, Barrett complained that too many contemporary poets were simply not taking their oracular duties seriously enough:

> In the eyes of the living generation, the poet is at once a richer and a poorer man than he used to be; he wears better broad-cloth, but speaks no more oracles: and the evil of this social incrustation over a great idea, is eating deeper and more fatally into our literature, than either readers or writers may apprehend fully. (Barrett Browning 1914, xiii)

This complaint is restated, in symbolic form, within the text of 'A Vision of Poets', in which a company of living poets refuse to sacrifice their lives and hearts for the cause of poetry. ('To give our hearts up? fie! that rage / Barbaric antedates the age; / It is not done on any stage.' [Barrett Browning 1914, 224].) Being unwilling to suffer for their art, they are denied the ability to speak 'oracles': unlike true poets (or true mesmerists), they have no access to visionary perception. Tellingly, when the spirits of the dead poets see this company of their unworthy would-be successors, they look at them as though they were insane:

> As a sane eye in master-passion
> Constrains a maniac to the fashion
> Of hideous maniac imitation
>
> In the least geste—the dropping low
> O' the lid, the wrinkling of the brow,
> Exaggerate with mock and mow,—
>
> So mastered was that company
> By the crowned vision utterly,
> Swayed to a maniac mockery. (Barrett Browning 1914, 224)

The 'sane eye', here, belongs to the dead poets, whose gaze masters these poseurs just as the quasi-mesmeric gaze of the madhouse-keeper was supposed to be able to master the lunatics under his charge (Shuttleworth 1996, 44). But what their gaze forces the living poets to do is not to behave sanely, but to *reveal their madness*, compelling them to engage in a 'hideous maniac imitation' of their predecessors which exposes their utter inability to emulate the poetic achievements of the true poets who have gone before them. In the eyes of society it is the true poets who seem insane, because of their willingness to throw away every worldly comfort in pursuit of a spiritual goal which brings them nothing but scorn and suffering; but the visionary perception of 'A Vision of Poets' reveals that it is the false poets who are the true madmen, for squandering their spiritual gifts for mere worldly advantage. Only a 'maniac', Barrett implies, would trade 'oracles' for 'better broad-cloth'.

Barrett's conception of poetry as a form of martyrdom via worldly neglect recalls Horne's 1833 diatribe on the sufferings of neglected men of genius, and some of the poem's more extreme claims—its assertion, for example, that even common labourers are happier and 'more approved' than poets, a claim which Horne also made in his 1833 *Exposition*—may owe something to his influence (Barrett Browning 1914, 216; Horne 1833, 12–13). Horne was a close friend of Barrett's at the time, with whom she had 'corresponded long and long' since 1839, although they had never met face to face. In the early 1840s she collaborated with him in writing a series of literary essays on contemporary writers, which were ultimately published in 1844 as *The New Spirit of the Age*, although Horne subsequently attempted to conceal the extent to which Barrett had contributed to the volume (Dally 1989, 80; Barrett Browning and Browning 1990, 28; Paroissien 1971, 274–81). However, while Horne had seen the neglect and mistreatment of 'men of genius' as a solvable social problem, Barrett describes it as an inevitable and glorious part of their destiny. For her, a true poet, like Tennyson's St Simeon, should be proud of the extremity of their suffering, because they know they suffer in a holy cause.

The essays which Barrett co-wrote with Horne for *The New Spirit of the Age* contain many additional examples of the exalted faith which both writers shared in the prophetic powers and duties of the true poet. The essay on Tennyson, for example, asserts that 'Poetic fire … has its source in the divine mysteries of our existence', and 'becomes the master passion of those who are possessed with it in the highest degree, and the most ennobling and refining influence that can be exercised upon the passions of others' (Horne 1844, 2:3). Poets 'of the highest class of genius', such as Tennyson, had to live an 'active life'; but the activity in which they needed to engage was all internal, involving 'inner struggles of the soul with its own thoughts' and 'dealings with the revelations which seem to come from other states of existence' (Horne 1844, 2:4, 2:29). The essay on Barrett herself, which was written by Horne, described her as writing 'like an inspired priestess', one 'whose individuality is cast upward in the divine afflatus, and dissolved and carried off in the recipient breath of angelic ministrants' (Horne 1844, 2:140). The essay on Carlyle restated Horne's old conviction that human progress was entirely due to the superior abilities of men of genius, and declared that 'the public mind' could never advance 'except by imbibing great principles from great men'—a category in which great writers and poets were clearly included (Horne 1844, 2:256).

The strongest defence of the poet as visionary in *The New Spirit of the Age* appeared in its final essay, on 'Henry Taylor and the Author of "Festus"', which seems to have been written primarily by Horne (Paroissien 1971, 280). This essay contested Taylor's 1834 critique of Shelley and 'the PHANTASTIC school', arguing that Taylor had been quite wrong to insist that poets should 'make all imagination subservient to the reasoning faculty, and what is called "good sense"' (Horne 1844, 2:288–9) Insisting that imagination was at least as important as judgement for the creation of great poetry, the essay reached an extraordinary crescendo in a passage which celebrated the 'Creative Passion' of true poets:

> There burns in the elements of certain natures, in the secret wells of their being, the deep sources where dwelleth the soul, a yearning towards some vaster region than the world which surrounds them, and an aspiration which would cleave its crystalline walls and soar away towards illimitable heavens, unknown ecstasies, and the eternal mysteries of Divinity. They feel this yearning, this aspiration, communicating itself to the very temperature and current of their blood; it stings them to the quick of inward being; it breaks out in drops upon the forehead, and rills down this poor inadequate, corporeal frame. They have mighty thoughts and deep; the deep thoughts often cross each other, and re-cross in their tumultuous lights and shades, till the man is vanquished by the cross-forces of his own mind: they see mighty phantasies and shapes; and the vision and the image rule over the man. (Horne 1844, 2:303)

This talk of the poet as one who is 'vanquished by the cross-forces of his own mind', and 'see[s] mighty phantasies and shapes' until 'the vision and image rule over the man', was exactly the sort of thing which had led Taylor—and, before him, Macaulay, Conolly, and Madden—to worry that the exercise of poetic imagination, unrestrained by reason, might ultimately plunge the would-be poet into insanity. (The same anxieties, as I have discussed, were raised by the growing popularity of mesmerism in the 1840s.) But Horne maintained that such a poet is *not* 'mad, or of intellect unsound', and that if he appears insane to ordinary people (which he probably will), it is only because 'he has awakened to more things than his fellows'. If he follows his 'Creative Passion', it will not take him to 'regions where he would be altogether strange and confounded' (i.e. madness); instead, it will lead him to the true spiritual 'freedom' which is his 'birthright' (Horne 1844, 2:303–4).

As I have discussed, the medical orthodoxy of the day regarded claims such as these with considerable scepticism. Horne viewed this suspicion of imaginative poetry as a cultural disaster, warning in *A New Spirit of the Age* that the British public were already much too willing to assume that poetry led to madness:

> Once convince and fortify John Bull in the opinion that to read poetry and cultivate his imaginative faculties will render him liable to aberration of mind, and it is all over with him, and the poets. He has half suspected this for a long time: his unsoundness is already on the other side. (Horne 1844, 2:43)

This may have been true of 'John Bull'; but it was emphatically not true of Elizabeth Barrett, who shared with Horne a view of the world in which the voice of the lone genius was always far more likely to be correct than that of conventional medical or critical 'good sense'. (She herself had experienced some bruising encounters with such medical orthodoxy, and would later describe to Browning how her doctor had tried to prevent her from writing entirely, as he 'gravely thought poetry a sort of disease' which was likely to bring about her early death [Felluga 2005, 121].) In 1845, expressing her amazement that a poet as great as Tennyson would rewrite his poems simply because he was told to do so by critics, she wrote:

> The curious thing in this world is not the stupidity, but the upper-handism of the stupidity. The geese are in the Capitol, and the Romans in the farm-yard—and it seems all quite natural that it should be so, both to geese & Romans! (Barrett Browning and Browning 1990, 21)

From such a perspective, any doctor or critic who mistook the inspired genius of the poet for madness was simply demonstrating that they were one of the geese rather than one of the Romans. As early as 1826 she had condemned the 'scientific fool[s]' who blamed madness on 'toiling intellect', insisting that 'too much learning maketh no man mad': and by 1844, after witnessing the powerlessness of a whole succession of doctors to cure or even identify her illness, she had more reason than ever to question the authority of the contemporary medical sciences (Barrett Browning 1914, 35–6).

By the time Martineau wrote to her about mesmerism in 1845, then, Barrett was strongly committed to a model of poets as visionary mystics. Her poetry had never contained any hint of the epistemological doubts which had dogged Tennyson and Browning since the early 1830s; for her, the vision of the poet was sacred, too holy to be questioned. (This may reflect the conditions in which she wrote: shut away in her sickroom, and making heavy use of morphine and ether, it is perhaps unsurprising that her visionary dream-world of poetic imagination should have come to seem more real to her as anything in the unvisited world outside [Dally 1989, 84–8].) A sense of the contrast between them can be gained by comparing Barrett's 'Vision of Poets' to part two of Browning's *Paracelsus*, of which it was in many ways a rewrite. In both works a poet who has sacrificed his life to the pursuit of beauty has a vision of the dead poets of the past, after which he dies. But whereas Browning's Aprile is a 'poor mad poet' whose pursuit of beauty without knowledge has damaged him so badly that he dies raving, convinced that he is a failure, Barrett's 'poet' is a figure of complete spiritual authority who dies convinced of his own success and immortality, having learned, like Keats, that the pursuit of beauty *is* the pursuit of knowledge, because beauty and truth are 'scarcely two'.

Similarly, both Barrett and Tennyson wrote poems in which self-torturing poets, scorned by the world, are comforted by angels who radiate 'glory' and speak mysterious words: but the contrast between them is instructive, and strongly representative of their respective approaches to such visionary encounters. Here is Barrett, in 'A Vision of Poets':

> The angel's smile grew more divine,
> The mortal speaking; ay, its shine
> Swelled fuller, like a choir-note fine,

Till the broad glory round his brow
Did vibrate with the light below;
But what he said, I do not know. (Barrett Browning 1914, 225)

And here is Tennyson, in poem 69 of *In Memoriam*:

They call'd me fool, they call'd me child:
    I found an angel of the night;
    The voice was low, the look was bright;
He look'd upon my crown and smiled:

He reach'd the glory of a hand,
    That seem'd to touch it into leaf:
    The voice was not the voice of grief,
The words were hard to understand. (Tennyson 1987, 2:384)

In Barrett's poem, the speaker does not know what the angel says because its words are reserved for the poet alone. But the poet *does* know, and what he hears fills him with joy and faith, so that despite the 'World's cruelty' he has suffered he is moved to prayer, declaring 'THOU, Poet-God, art great and good!' (Barrett Browning 1914, 226–7). In Tennyson's version of the same theme, however, even the poet who hears the voice of the angel cannot understand its message. All of Barrett's statements are definite ('The angel's smile grew more divine'), while Tennyson's are uncertain and tentative: when the 'angel of the night' touches the speaker's crown of thorns he only '*seem'd* to touch it into leaf' (my italics), and his statement that the angel's voice is 'not the voice of grief' tells us almost nothing about what it *is* the voice of. (The answer could be 'hope' or 'faith'—but it could just as easily be 'delusion' or 'despair'.) At every point, Tennyson's poem leaves open the possibility that the speaker really *is* the 'fool' that 'they' think he is, whereas Barrett's calls in no less an authority than 'THOU, Poet-God' to aggressively foreclose all interpretations other than the one that she intends.

I would suggest that Barrett's views on mesmerism (and, later, spiritualism) grew out of her insistence on the authoritative validity of visionary perception, especially when the message of such visionaries coincided with the teachings of Christianity: she had argued in 1838 that all true poets naturally gravitated towards spiritual and religious subject matter, because their pursuit of truth would inevitably lead them away from worldly things and towards the divine (Barrett Browning 1914, 76–8). That she saw a strong resemblance between the trance states induced by mesmerism and the visionary perception of the inspired poet is clear from her later epic *Aurora Leigh* (1856), in which she speculated upon why 'nearly all young poets should write old':

> It may be, perhaps,
> Such have not settled long and deep enough
> In trance, to attain to clairvoyance,—and still
> The memory mixes with the vision, spoils,
> And works it turbid. (Barrett Browning 1914, 365)

In 1856, 'trance' and 'clairvoyance' were still technical terms, seldom used except in the context of mesmeric and spiritualist phenomena. For mesmerists and poets alike, she implied, it takes time and practice to attain the true 'clairvoyance' of visionary perception—the one exception being the 'strong excepted soul' of that arch-Romantic, John Keats. For the immature poet, as for the novice clairvoyant, the 'vision' of transcendent spiritual realities tended to be corrupted by an admixture of mere 'memory' of the things of this world—what Horne's and Barrett's essays variously referred to as the opposing forces of 'ideality' and 'reality', 'spirit' and 'matter', or 'the ideal' and 'the real' (Horne 1844, 2:9, 2:19, 2:155–6, 2:180, 2:260). But if one *did* 'settle ... deep enough / In trance', working patiently to cultivate one's poetic and/or mesmeric powers, then the visionary insights that resulted would be real and true and eternal, imparting both to the seer and to their enlightened audience a faith which no amount of popular scepticism or institutional disapproval should ever be allowed to shake. Elsewhere in *Aurora Leigh*, she wrote of how, in 'the artist's ecstasy', 'The spiritual significance burn[s] through / The hieroglyphic of material shows': terms which recall not only the pronouncements of Wordsworth and Shelley on the role of the poet-as-prophet, but also Martineau's descriptions of her experiences during mesmeric trances a decade before (Barrett Browning 1914, 467). In consequence, for Barrett, as for Horne and Townshend, faith in mesmerism and faith in visionary poetry went very much hand in hand.

'Super-Belief': Mesmerism in the Barrett-Browning Courtship Correspondence

The year in which Martineau wrote to Barrett about mesmeric therapy, 1845, was also the year in which Barrett and Browning began their surreptitious courtship, and mesmerism was one of the subjects that they discussed over the course of their ensuing covert correspondence. In keeping with their respective attitudes towards visionary inspiration in general, Barrett was enthusiastic about the possibilities of the new science, while Browning was considerably more sceptical (Karlin 1989, 66–7). Mesmerism, Barrett wrote playfully to Browning, was something for which 'no 'theory' could account': 'Are you not aware that these are the days of Mesmerism & clairvoyance? are you an infidel?' When Browning complained to Barrett that one of his friends believed in mesmerism, 'as in all other unbelief', she wrote back: 'Bur belief in mesmerism is not the same thing as general unbelief—to do it justice—now is it? It may be super-belief as well' (Barrett Browning and Browning 1990, 8, 28, 77, 79).

'Unbelief' and 'super-belief' were of special importance to Barrett during this correspondence, as she constantly struggled with the question of whether she could allow herself to believe in the reality of Browning's love for her. Requesting him to send her a lock of his hair as a love-token, she wrote: 'You have come to me as a dream comes, as the best dreams come… dearest—& so there is need to me of 'a sign' to know the difference between dream & vision'. A few weeks later she wrote again, comparing her situation to that of a traveller coming unexpectedly across a fountain in the desert: 'to doubt whether it may not all be *mirage*, would be the natural first thought … the recurring dream-fear! … now would it not?' (Barrett Browning and Browning 1990, 160, 168). Her worry was not that Browning might be deliberately deceiving her, but that he might have unintentionally deceived *himself* by falling in love with an imaginary version of her which existed only in his own mind, and which could not survive contact with reality. As she wrote:

> [Browning's visits] confounded me, & left me constantly under such an impression of its being all dream-work on his side, that I have stamped my feet on this floor with impatience to think of having to wait so many hours before the 'candid' closing letter c[d] come with its confession of an illusion. 'People say,' I used to think, 'that women *always* know … & certainly I do not know … & therefore … therefore'—The logic crushed on like Juggernaut's car. (Barrett Browning and Browning 1990, 182–3)

Horne had described Barrett as an 'inspired priestess': but whatever visionary certainties her 'inspired' poetic role might have granted her, insight into Browning's mind was evidently not among them. In its absence, she had to take it on trust that his appearance in her life was a true 'vision' rather than a 'dream' or a '*mirage*'. Alluding to the classical myth that all dreams come to the world through either the gate of horn (if true) or the gate of ivory (if false), she wrote to Browning:

> [Y]ou wonder at me for looking twice, thrice, four times, to see if it [i.e. his love for her] comes through ivory or horn. You wonder that it should seem to me at first all illusion—illusion for you, ... illusion for me as a consequence. But how natural— (Barrett Browning and Browning 1990, 175)

Faced with this dilemma, Barrett resolved to follow the same course in love that she had previously taken with religion, mesmerism, and poetry, and to reject 'unbelief' for 'super-belief'. Thus her declaration to Browning in December 1845 that 'It is with me as with the theologians': 'Surely I believe in you and in "mysteries". Surely I prefer the no-reason to ever so much rationalism. (rationalism & infidelity go together they say!)' Yet she reserved the right to 'be afraid sometimes notwithstanding', and an undercurrent of anxiety surely lurked beneath her insistence that he accept her gift of an ivory pen-holder, 'made from a splinter of the ivory gate of old', thus connecting his writings—including, presumably, his love letters to her—with the 'ivory gate' of dreams that turn out to be false rather than true (Barrett Browning and Browning 1990, 173–4, 199).

For Barrett, belief in love, poetry, and mesmerism all went together: all were a matter of believing in 'mysteries', whether or not one understood them. Browning insisted that there was no resemblance between his love for Barrett and the mysteries of mesmerism, with the former being straightforwardly true and the latter straightforwardly false, but disentangling the connections between mesmerism and poetry was not so easily accomplished. As I have mentioned, in the 1830s and 1840s, mesmerism was repeatedly associated with Wordsworthian poetic vision: explicitly by Colquhoun and Townshend and implicitly by Barrett and Martineau. After their marriage, Robert and Elizabeth Barrett Browning moved to Italy, and in 1847 they settled in Florence; there they became friends with Tennyson's brother Frederick, who was himself a poet, mesmerist, and Swedenborgian, and saw his poetic and mesmeric activities as being entirely complementary (Dally 1989, 158; Thomas 1982, 149). Under

Frederick Tennyson's influence, Barrett Browning became increasingly involved with spiritualism, much to Browning's disapproval. And yet, as Karlin points out, when Browning unsuccessfully attempted to rewrite *Sordello* in the 1850s, he explicitly described Sordello's poetic power in mesmeric terms, and depicted the medieval poet as speculating anachronistically about 'the mesmeric effort I can command' (Karlin 1989, 77 n 15).

This image of the poet as mesmerist recalls the claims made by Macaulay in 1825 regarding the near-supernatural power with which poetry could affect the young, the ignorant, and the primitive, in all of whom it could produce 'plenitude of belief'. As he wrote, 'Poetry produces an illusion on the eye of the mind, as a magic lantern produces an illusion on the eye of the body' (Macaulay 1843, 9). For Macaulay, poets were thus engaged in the same activity of which mesmerists and spiritualists were accused by their critics: that of persuading people, themselves often included, to regard illusions as realities. This, as I have discussed, was exactly what Barrett had feared that Browning was doing during their courtship correspondence: first deluding himself with a fantasy-image of her, and then using the power of his imagination to compel her to accept that fantasy as the truth despite its lack of any firm foundation in reality, just as Colquhoun argued that phreno-magnetists were doing to themselves and their patients. In person his presence could be quite as overpowering as that of any mesmerist, as Barrett explained to him in a letter of 1846. 'You *influenced* me', she told him, 'in a way in which no one else did':

> Then, when you came, you never went away—I mean I had a sense of your presence constantly. Yes … & to prove how free that feeling was from the remotest presentiment of what has occurred. I said to Papa in my unconsciousness the next morning … 'it is most extraordinary how the idea of Mr. Browning does beset me—I suppose it is not being used to see strangers, in some degree—but it haunts me … it is a persecution.'
>
> […]
>
> Do you know that all that time I was frightened of you? frightened in this way. I felt as if you had a power over me & meant to use it, & that I could not breathe or speak very differently from what you chose to make me. As to my thoughts I had it in my head somehow that you read *them* as you read the newspaper—examined them & fastened them down writhing under your long entomological pins—ah, do you remember the entomology of it all? (Barrett Browning and Browning 1990, 217)

As Winter notes, Barrett describes Browning here 'in strikingly mesmeric terms' (Winter 1998, 239). Browning was evidently uncomfortable with Barrett's description of him as an overmastering, mesmerically powerful figure, able to read her thoughts and control her actions even when he was no longer physically present, and he repeatedly insisted that she held at least as much power in their relationship as he did. As he wrote to her early in their correspondence, 'I really believe you to be my superior in many respects, and feel uncomfortable till *you* see that, too' (Barrett Browning and Browning 1990, 59–60).

Barrett, however, was clearly correct in at least one respect. Browning may have been in love with a real woman rather than a fantasy, but during their courtship, he really *was* fond of imagining a fantasy-image of her sitting beside him, while the real Elizabeth Barrett was absent: as he wrote, '*always* I turn in the dusk and *see* the dearest real Ba with me'. When she learned of this, Barrett reproached him with contenting himself with fantasies of her, while neglecting the reality: 'The ἀϊδωλου sits by the fire—the real Ba is cold at heart through wanting her letter' (Barrett Browning and Browning 1990, 233). In response to Barrett's persistent worries that he was only in love with an imaginary dream version of her, Browning stressed that he understood the difference between fantasies and realities, real women and bodiless illusions:

> You *know* by this that it is no shadowy image of you and *not* you, which having attached myself to in the first instance, I afterwards compelled my fancy to see reproduced, so to speak, with tolerable exactness to the original idea, in you, the dearest real *you* I am blessed with—you *know* what the eyes are to me, and the lips and the hair— (Barrett Browning and Browning 1990, 230)

Yet as the plans for their secret marriage and subsequent elopement progressed, Browning must have known that he was in danger of appearing to be a man who had done more or less what contemporary writers were constantly worrying about unscrupulous mesmerists doing: using his superior mental talents and imaginative power first to mesmerise a woman into loving him ('it haunts me ... it is a persecution'), and then to summon her away from her father's house and compel her to run away with him, effectively transforming his mental image of 'the dearest real Ba with me' into a physical reality through the sheer mesmeric power of his personality. This uncomfortable similarity between the mesmerists whom Browning condemned and Browning himself, both as a poet and as a lover, hovers uneasily beneath the surface of one of Browning's most disturbing poems, 'Mesmerism'—a poem which he published in 1855, the same year as his disastrous séance with Home.

'All I Believed Is True': Browning's 'Mesmerism'

'Mesmerism' was not published until 1855, although Winter has speculated that it may have been written as early as 1845 (Winter 1998, 240–1). Rebuking the mesmeric enthusiasms of writers such as Townshend, it expresses Browning's intense discomfort at the idea of deliberately granting the imagination priority over reality, and of assuming that the mere vividness and intensity of one's mental experiences could ever constitute evidence of their actual validity.

The speaker of 'Mesmerism' is a mesmerist who has decided to use his mesmeric powers in order to exert mental control over a woman, summoning her into his presence and compelling her to do his bidding, 'as befits a thrall'—exactly as Barrett had feared that Browning might be doing to her. In choosing this scenario, Browning was tapping into a strong tradition of anti-mesmeric scaremongering: mesmerists tended to be older men, while the 'clairvoyants', 'somnambulists', and 'ecstatics' whom they 'magnetised' and mesmerised were often younger and/or lower-class women, leading to much prurient speculation as to what other ends the apparently limitless power which mesmerists wielded over their female patients might be turned to (Ruth 1999, 313–4; Winter 1998, 101–2). The definitive version of this story, in which a young woman is mesmerised and subsequently exploited by an older man, would appear in 1894 in the form of George Du Maurier's *Trilby*: as Auerbach notes, Stoker's *Dracula* is, in many respects, only a minor variation on the same theme (Auerbach 1982, 16). In 'Mesmerism', however, Browning focused not only upon the moral dangers of mesmerism, but also upon the epistemological uncertainties to which it might give rise.

As with so many of Browning's dramatic monologues, it is the rhythms of 'Mesmerism' which provide the first hint that its speaker is not quite in his right mind. It begins:

All I believed is true!
  I am able yet
  All I want, to get
By a method as strange as new:
Dare I trust the same to you?

If at night, when doors are shut,
  And the wood-worm picks,
  And the death-watch ticks,

And the bar has a flag of smut,
And a cat's in the water-butt—

And the socket floats and flares,
   And the house-beams groan,
   And a foot unknown
Is surmised on the garret-stairs,
And the locks slip unawares— (Browning 1983–2009, 5:105–6)

The primary metre is trochaic, which in English lends itself to an unnatural, chant-like rhythm, most famously exhibited in the nineteenth century by Longfellow's *Hiawatha*; but here each trochaic line ends with a stressed hyperbeat, so that rather than dying sonorously away like Longfellow's, they jerk upwards with a sort of nervous twitch. As in 'Porphyria's Lover' and 'Johannes Agricola', the uneasy effect of these lines is enhanced by their unusual rhyme scheme: the earlier poems had used an ABABB rhyme scheme, whereas 'Mesmerism' uses an ABBAA pattern, but the result in both cases is to create an effect of imbalance, as though their speakers are aiming for a more conventional ABAB rhyme scheme but can't quite bring themselves to let go of their thoughts at the appropriate time. Even more than in 'Porphyria's Lover', the extreme brevity of the lines in 'Mesmerism' means that, for a reader accustomed to the rhythms of English pentameter, the rhyme words seem to come too close together, like someone obsessively circling back to the same ideas; this is particularly the case with the two B-rhymes, which in each stanza come just five syllables apart. Form and metre alike are slightly irregular, so that no two stanzas are quite the same, preventing the reader from settling into any kind of predictable pattern; the poem remains uncomfortable, always slightly off-kilter. The reason for the speaker's disturbed mental state becomes apparent when he describes what exactly he has been doing 'at night, when doors are shut':

If since eve drew in, I say,
   I have sat and brought
   (So to speak) my thought
To bear on the woman away,
Till I felt my hair turn grey—

Till I seemed to have and hold,
   In the vacancy

  'Twixt the wall and me,
From the hair-plait's chestnut gold
To the foot in its muslin fold—

Have and hold, then and there,
  Her, from head to foot,
  Breathing and mute,
Passive and yet aware,
In the grasp of my steady stare— (Browning 1983–2009, 5:106–7)

The crucial word, here, is 'seemed'. The mesmerist tries to call 'the woman away' into his presence by imagining that she is already there: he has pictured her, in obsessive detail, until she 'seemed' to actually stand in the room with him. He claims that this mesmeric call has the power to actually bring her to him, so that his imagined image of her is replaced by the physical reality of her presence:

On doth she march and on
  To the fancied shape;
  It is, past escape,
Herself, now: the dream is done
And the shadow and she are one. (Browning 1983–2009, 5:111)

By the end of the poem, the mesmerist has convinced himself that his mesmeric summoning has succeeded, and that the woman he has called is now present in the room, compelled by his mesmeric powers to take the place of his 'fancied shape' and do whatever he commands. But is this what has actually occurred? He has 'surmised' a 'foot unknown' outside his room—but is his surmise actually correct? Has he really called the woman to him, or has he simply become so lost in his own fantasies that he can no longer distinguish between imagination and reality?

Being a dramatic monologue, the poem gives us no information on what is 'really' happening: it only tells us what the rather unhinged speaker *thinks* is happening. The very essence of his mesmeric craft consists of imagining things with such intensity that they 'seem' to be really present, and thus risks leaving him unable to distinguish between fantasies that are only imagined and fantasies that are actually fulfilled. If 'the fancied shape' is 'herself, now', and 'the shadow and she are one', it may be because the real woman has entered the room and taken the place of the imaginary 'shadow', as the real Elizabeth Barrett ultimately replaced Browning's fantasy-image of 'the dearest real Ba with me'. But it may also be because the speaker no longer has any way of reliably telling the two apart.

Hints of the mesmerist's epistemological dilemma multiply in the later stanzas. By summoning the woman, he asserts that he shall:

> Command her soul to advance
> And inform the shape
> Which has made escape
> And before my countenance
> Answers me glance for glance— (Browning 1983–2009, 5:108)

In this stanza, the mesmerist still recognises that the 'shape' before him is imaginary, requiring a real body to 'inform' it; but the imaginary woman before him no longer seems to be under his control. It has done precisely what the real woman is supposed to be unable to do, namely, 'made escape'; it now seems to him to be something external to himself, standing before him and 'answer[ing] [him] glance for glance', like the phantom-woman involuntarily conjured by the speaker's imagination in Tennyson's 'Oh! That 'Twere Possible'. When he believes that the real woman has arrived, Browning's speaker asks himself: 'Have I drawn or no / Life to that lip?', by which he probably means that it may be his mesmeric power which has 'drawn' blood into her lips, making her look glowingly alive. But another reading is possible: he may have 'drawn' the life into the lip he imagines in the same way an artist might draw it on paper, sketching it in his mind's eye rather than actually conjuring it in reality. He declares that he '*must* believe' (my italics) that his summoning is efficacious; so when, later in the poem, he claims that 'I see my belief come true', it is not at all clear that this is anything more than a wish-fulfilment fantasy, fuelled by the desperate force of his obsessive desire.

What 'Mesmerism' brilliantly illustrates is the ease with which imagination can lapse into solipsism or delusion, allowing people to believe that they were employing the kind of deliberate, active imagination praised by Colquhoun (and by Coleridge) while they were in fact prisoners of the kind of passive, reactive imagination feared by Forbes. The practitioner and subject of mesmerism could easily be seen as enacting a kind of *folie à deux*, with the mesmerist imagining that they feels a magnetic force flowing from them into their patient, and the patient imagining that they feel the same (purely imaginary) force affecting their body or mind. Spiritualist mesmerism was an even easier target: for when the entire process consists of finding someone unusually receptive to suggestion and then encouraging them to call spirits, angels, and, so on, before their mind's eye, there is surely a very strong chance that what they go on to describe will not be an authentic vision of the spirit world, but simply a range of images drawn from their own imagination at the prompting of the presiding mesmerist, which mesmerist and clairvoyant alike then agree to regard as realities.

In the last stanzas of 'Mesmerism', Browning's speaker turns to address God:

> First I will pray. Do Thou
> That ownest the soul,
> Yet wilt grant control
> To another, nor disallow
> For a time, restrain me now!
>
> I admonish me while I may,
> Not to squander guilt,
> Since require Thou wilt
> At my hand its price one day
> What the price is, who can say? (Browning 1983–2009, 5:111–12)

The idea that there was something impious, even Satanic, about the power which mesmerists wielded over their patients was frequently expressed in the 1840s: Townshend mentioned this belief in his works on mesmerism, and another of Basil Montagu's friends, the clergyman George Sandby, devoted a large part of his book *Mesmerism and Its Opponents* (1844) to rebutting the claim that mesmerists drew their power from the Devil (Townshend 1854, 108; Sandby 1844, 4–26; Winter 1998, 260–2; Fulford 2004, 70–1). To many anti-mesmerists it seemed incredible that God should grant one human being the ability to apparently extinguish the free will of another, and in these lines, Browning's speaker hardly seems to be able to believe it himself. His prayer, 'Do Thou … restrain me now!' seems to be half-way between a demand (God *must* restrain me) and a question (*does* God restrain me?), and he ultimately 'admonish[es]' himself 'not to squander guilt': God will one day make him pay the 'price' for this misuse of his mesmeric powers, so he might as well make the most of it while he has the chance, presumably by raping the woman whom he believes that he has mesmerically summoned to his room. It is a backflip of moral reasoning worthy of Browning's 'Johannes Agricola', and it also serves to link him with Johannes' fellow-inmate in the 1842 'Madhouse Cells', Porphyria's lover—a man who, like the mesmerist, sits looking at another silent, immobile woman, wondering at the fact that 'God has not said a word!' When the mesmerist asks, rhetorically, 'What the price is, who can say?', he *means* that God will one day judge him for using his abilities in this way, and that what this judgement will be is, like all of God's judgements, unknowable. But if his mesmeric powers are merely delusions induced by his overpowering imagination, then the reader is in a position to answer his final, rhetorical question: the price is madness, and he is already paying it.

'Mad—Inspired—Say Either': Browning's 'Mesmerism', Barrett's 'Lady Geraldine's Courtship', and the Séance of 1855

As Browning's rewrite of *Sordello* to refer to the poet's 'mesmeric effort' suggests, the risks that Browning evokes in 'Mesmerism', of becoming so entranced by the power of one's own imagination that one loses touch with morality and reality, are also applicable to poetry. Browning's mesmerist functions as a nightmare image of a poet who uses his ability to make himself and others imagine things for purely selfish ends, possibly at the cost of his own sanity. As I have suggested above, the poem also functions as a nightmarish rewriting of Browning and Barrett's courtship, in which all her anxieties turn out to be true and the entire 'relationship' is simply a one-sided charade choreographed by a solipsistic lunatic. By expressing his disquiet at the mental and moral risks involved in the practice of mesmerism via a poem in which a man imagines a woman's presence so intensely that reality seems compelled to follow suit, Browning was drawing upon contemporary anxieties about the ways in which mesmerism could be used to manipulate the behaviour of impressionable young women. But he was also rewriting a scene from one of Barrett's most popular poems: the widely praised 'Lady Geraldine's Courtship', which had been published in 1844.

'Lady Geraldine's Courtship' held a special significance for Browning, as it contained lines praising his own poetry. It was by reading these lines that Browning first learned that Barrett held his work in high regard, and it was this knowledge, in turn, which in 1845 encouraged him to begin the correspondence with her that ultimately led to their elopement and marriage (Thomas 1982, 102–3). But its relevance to 'Mesmerism' is due to its 'Conclusion', in which a low-born poet, Bertram, sees what he thinks is a vision of his aristocratic beloved, Geraldine, with whom he believes he has ruined his relationship by subjecting her to a furious harangue in front of her friends:

> Soh! how still the lady standeth! 'tis a dream!—a dream of mercies!
> 'Twixt the purple lattice-curtains, how she standeth still and pale!
> 'Tis a vision, sure, of mercies, sent to soften his self-curses—
> Sent to sweep a patient quiet o'er the tossing of his wail. (Barrett Browning 1914, 214)

Bertram believes that he is hallucinating, and that what he sees is only a 'Vision of a lady', a 'dream of Geraldine'. But, in fact, she turns out to be really present: and the passionate outburst to which he subjected her earlier in the day, far from ending their relationship, has actually won her love. During that outburst, Bertram had felt himself to be both inspired and insane, described himself as being 'mad—inspired—say either!'; in his frenzy he had felt himself dilated 'to a Pythian height', able to see angels in the sky and to '[know] what spirits can'. His rage had made Geraldine tremble, because 'In the presence of true spirits', this is all that 'a worldly man or woman' is capable of doing: 'what else *can* they do but quail?' In this state he had 'spake out wildly, fiercely, brutal truths of her and others' until finally his 'strength of passion' makes him faint (Barrett Browning 1914, 211–12). Unsurprisingly, he had assumed that following such 'brutal' behaviour he would never see her again: but instead it turns out that she has been completely persuaded by all the 'inspired' things he said to her, and is now happy to marry him despite the difference in their social status.

In 'Lady Geraldine's Courtship', then, the fear that the poet might be mad is raised only in order that it might be dispelled by the revelation that he is in fact ultra-sane—saner, in fact, than the 'worldly' people around him. (Recall Barrett and Horne's praise of the true poet in *A New Spirit of the Age*: 'Does he dream? No, he wakes; he has awakened to more things than his fellows.') (Horne 1844, 2:303). What Bertram thinks is an involuntary hallucination of Geraldine turns out to be an accurate vision of her real presence, and what he thinks is a mad and 'brutal' outburst of socially unacceptable behaviour turns out to be an inspired speech capable of winning over the woman who hears it. His actions are endorsed, his inspiration confirmed, his imagination validated, and his authority bolstered—all very much at Geraldine's expense. Crucially, while the body of 'Lady Geraldine's Courtship' is written in the form of a letter that the poet writes to his friend (and is, thus, like Browning's dramatic monologues, subjective and potentially unreliable), the 'Conclusion' in which the reality of his vision is revealed is narrated in the third person, thus confirming the authenticity of his sanity and his power. Earlier in the poem Bertram insists that even if the world was wrapped in the 'hot electric breath' of the telegraph, it would comprise 'no new spirit-power'. But the inspiration of the poet, like the 'electric breath' of the mesmerists in which Barrett had 'long been a believer', really *does* bring a 'new spirit-power' to the world—a power which proves sufficient to bring the humbled Geraldine to Bertram's side (Barrett Browning 1914, 210).

Browning's 'Mesmerism' is, among other things, a takedown of these sorts of pretensions, so dear to Horne and Townshend, in which one's actions are always justified and one's perceptions are always accurate because of the special kind of mesmerically and/or poetically gifted person one believes oneself to be. In 'Mesmerism', the speaker believes that the power of his mesmeric mind can compel the woman to come to him and fulfil his desires, just as the power of Bertram's 'inspired' genius really *does* compel Geraldine to come to him and marry him in 'Lady Geraldine's Courtship'. But whereas 'Lady Geraldine's Courtship' insists that the poet really *is* the kind of superior being he thinks he is, 'Mesmerism' strongly suggests that the mesmerist is simply lecherous, egotistical, and possibly insane: all traits which would probably also have been attributed to Bertram by the 'worldly' men and women who witness his 'mad—inspired—' outburst in front of Geraldine.

I would suggest that in their opposite reactions to this scenario, in which a supposedly mentally superior man apparently compels a woman to submit to him through the sheer power of his mental talents, can be seen the germ of all of Browning and Barrett's subsequent disagreements over mesmerism and spiritualism, from their correspondence about Martineau in 1845 to their opposite reactions to Home's séance a decade later. Building upon the critique he had developed in *Paracelsus* and *Sordello*, Browning viewed such claims to special power and insight as being both morally and epistemologically suspect; but for Barrett, who was strongly attached to the belief that visionary perception won through suffering was much more likely to be valid than the judgements of the 'worldly', the voice of the socially marginalised seer—whether poet, mesmerist, or spiritualist—always had to be taken seriously. The fact that their claims might seem doubtful or even absurd to less spiritual individuals did not, for her, constitute a valid objection. After all, as she wrote in 'A Lay of the Early Rose' (1844), 'What angel but would seem / To sensual eyes, ghost-dim?' (Barrett Browning 1914, 262).

The disagreement between Barrett and Browning over the validity of mesmerism was a portent of the much more severe disagreement over spiritualism which would come between them in the 1850s. Through her friendships with Frederick Tennyson and Robert Lytton in Florence, Barrett—now Mrs Barrett Browning—swiftly became immersed in spiritualist culture in Italy, attending regular séances and encouraging her maid, Wilson, to experiment with spirit writing. Browning, for his part, was alarmed by her ever-deepening involvement in practices which he regarded as a compound of charlatanry and delusion, and feared that they might have a negative impact upon her mental health. Barrett Browning was disappointed by Browning's refusal to believe in spiritualism—as she complained to her brother, George, in 1853, 'Really he is dying hard'—and when they visited London in 1855, with Home at the height of his fame, she was eager to take part in one of his séances (Kelley et al. 1984–2016, 19:163–71). True to form, however, Browning took against Home at once, judging his behaviour to be full of 'unmanlinesses' that were 'in the worst taste' (Dally 1989, 158–61, 167; Thomas 1982, 149, 156–9).

What happened in the séance that followed is deeply unclear, because the three principal people involved—Browning, Barrett Browning, and Home himself—all gave very different accounts. According to Barrett Browning, spiritual hands appeared, played musical instruments, turned and then raised the table, and finally placed a wreath of clematis on her head: a series of supernatural manifestations which she regarded as a 'wonderful and conclusive' proof of the validity of spiritualism. According to Browning, the whole thing was 'a cheat and an imposture': the 'hands' were simply props which Home manipulated with his feet, and at one point he had actually caught hold of the medium's foot under the table while he was engaged in just such a trick. According to Home, Browning's objections were fabrications motivated by rage and envy: he had hoped that the spirits would place the wreath on *his* head, and 'seemed much disappointed' when they crowned his wife with it instead. Home claimed that this slight, coupled with his inability to explain what he had just witnessed, made Browning so angry that at their next meeting he became 'pallid with rage' and behaved like a 'maniac'. Home and Browning finally parted on furiously bad terms. Caught in the middle, Barrett Browning could only apologise, miserably repeating: 'Dear Mr Home, I am not to blame. Oh dear! Oh dear!' (Dally 1989, 167–8; Thomas 1982, 157–9).

Browning would ultimately have his revenge upon Home by caricaturing him as a hypocritical charlatan in his poem 'Mr Sludge, the Medium'—although, perhaps in deference to his wife's feelings on the matter, he did not publish the poem until after her death. But their opposite reactions to the séance serve as a dramatic example of their opposing views on all matters of visionary experience. Barrett Browning always wanted to believe that the spirits and visions she heard of were real ones, and that those who saw them were telling the truth: thus her faith in the authority of mesmerists and mediums, which stemmed from the same source as her faith in visionary poets. She was drawn to Browning himself, despite her fears that his love for her might not be in earnest, precisely because she saw him as just such a figure of irresistible visionary power: as she wrote to him early in their courtship correspondence, 'You are Paracelsus—and I am a recluse with nerves that have been all broken on the rack', once again positioning Browning as the mesmerically authoritative magus and herself as the passive patient whom he could command at will (Barrett Browning and Browning 1990, 33). But Barrett seems to have seen Browning as being like Paracelsus in the earlier parts of Browning's poem, when he is still full of confidence in his own power and mastery, whereas Browning himself seems to have identified much more with the disillusioned Paracelsus of the poem's later sections, no longer believing in his own ability (or that of any other single individual) to change the world through the sheer exercise of mental force.

For his part, Browning admired Barrett Browning for her ability to express her own deepest thoughts and feelings in her poems in a way that he did not, and increasingly did not even attempt to do, preferring the shadow-play of his endlessly questionable dramatic monologues. He regarded her as a greater poet than himself specifically because of her capacity to articulate her own personal vision of life in her poems—a capacity which was at least partly due to her belief that the true poet must be a fearless preacher of visionary truths to the world, whose 'office' was 'the destruction of the conventions of the hour', very much like the mesmerists and spiritualists whose claims she was so eager to believe (Barrett Browning and Browning 1990, 15, 26). Yet in matters of mesmerism and spiritualism, as in matters of poetry, Browning's intense wariness of people who claimed to possess special forms of power or insight repeatedly put him on the side of the sceptics. Like the man who frankly told Martineau that even if he witnessed the phenomena of mesmerism with his own eyes, 'he would sooner think himself and the whole company mad than admit them', Browning's hostility to these movements was so intense that *he* risked being the one who seemed irrational or insane, as Home's comparison of him to a 'maniac' implied (Martineau 1845, 3). Yet, for him, the stakes were high: for if the mesmerists and spiritualists were correct, and transcendent visionary insight really *was* within easy reach of every person of superior mental abilities, then his abandonment of his early Shelleyan enthusiasm would represent no less than a dereliction of his spiritual duties as a poet.

Conclusion: Faith and Doubt

As I have argued, the second quarter of the nineteenth century witnessed the development of a new 'medico-psychological' orthodoxy in Britain, which explained poetic talent in terms of the abnormal development of the brain and nervous system. The mesmeric medicine of the period concurred in interpreting poetic talent in basically physical terms: for mesmerists such as Townshend, poets possessed the gifts that they did because their nerves had a superior ability to channel the mysterious magnetic fluid upon which mesmeric theory was based (Townshend 1821, 83). But while mainstream medical writers such as Browne, Conolly, Winslow, and Madden saw poetic imagination as being hazardous to sanity unless it was strictly moderated, counter-balanced, and controlled, pro-mesmeric writers such as Sandby and Colquhoun were willing to credit the imagination with a much more active, positive force, identifying it with that mesmeric power through which the truly 'magnetic' man was able to influence the world (Colquhoun 1836, 1:141–53, 1:165–9). As a result, such mesmeric theories were popular with many authors of the period, and especially with those such as Barrett, Horne, and Townshend who still remained loyal to the Shelleyan ideal of the poet as a mystic visionary, able to perceive higher realities hidden from ordinary men. For them, mesmerism provided an apparently scientific framework for explaining why poets possessed such abilities, and why the visions and intuitions of the 'inspired' poet deserved to be taken seriously as genuine insights, rather than dismissed as signs of incipient insanity produced by their malfunctioning brains.

Given that virtually every other author of the period, from Dickens to Tennyson to Bulwer-Lytton, seems to have at least flirted with mesmerism,

Browning's stalwart scepticism stands out as highly unusual. I would suggest that this scepticism is best understood in relation to his hardening insistence that the task of poetry, or at least of *his* poetry, was to look *at* the world rather than through it: an insistence which brought with it that mounting horror of imaginative solipsism which resonates through his poetry from *Paracelsus* in 1835 to 'Mesmerism' in 1855. Mesmerism, with its insistence that imagination can become reality if only we imagine hard enough, came worryingly close to the creeds that made characters like Aprile and Sordello so ineffectual, and those like Johannes Agricola and Porphyria's Lover so dangerous. While writers such as Horne and Townshend indulged in fantasies about the spiritual superiority of their own genius, Browning continued to warn that such fantasies could lead to very dark places, and psychologically scrutinised them in his poetry with the same ruthlessness with which the mad-doctors of the period evaluated their patients.

By the 1850s, however, this had become a contentious opinion among British poets. Wordsworth, Keats, and Shelley, who had been little read in 1830, had by 1850 become touchstones to a whole new generation of would-be poetic visionaries: and just as spiritualism responded to the mid-Victorian crisis of faith, which Tennyson had so heartbreakingly evoked in *In Memoriam*, by staging ever-more *outré* displays of supernatural 'phenomena', so mid-century poets such as Sydney Dobell and John Stanyan Bigg responded to the doubts which had gathered around the ideal of visionary poetic inspiration since 1825 by competing to demonstrate the extent of their visionary prowess in verse. It is this mid-century resurgence of Romantic enthusiasm, and the responses which it drew from Browning and Tennyson which forms the subject of my final chapter.

References

Primary Sources

Barrett Browning, Elizabeth. 1914. In *Poetical Works*, ed. Frederic Kenyon. London: John Murray.

Barrett Browning, Elizabeth, and Robert Browning. 1990. In *The Courtship Correspondence 1845–6*, ed. Daniel Karlin. Oxford: Oxford University Press.

Blake, William. 1988. *Complete Poetry and Prose*, ed. David Erdman, revised edition. New York: Anchor.

Browning, Robert. 1983–2009. *Poetical Works*, eds. Ian Jack, Rowena Fowler, Margaret Smith, Robert Inglesfield, et al., 15 vols. Oxford: Clarendon.

Cahagnet, Louis Alphonse. 1851. *The Celestial Telegraph; or Secrets of the Life to Come, Revealed Through Magnetism*. New York: J.S. Redfield.
Coleridge, S.T. 1983. *Biographia Literaria*, eds. James Engell and W. Jackson Bate, 2 vols. Princeton: Princeton University Press.
———. 1999. In *Poems*, ed. John Beer. London: David Campbell.
Colquhoun, J.C. 1836. *Isis Revelata*. 2 vols. Edinburgh: Maclachlan and Stewart.
———. 1843. *The Fallacy of Phreno-Magnetism Detected and Exposed*. Edinburgh: William Wilson.
Crowe, Catherine. 1848. *The Night Side of Nature*. 2 vols. London: T.C. Newby.
Forbes, John. 1845. *Mesmerism True – Mesmerism False*. London: John Churchill.
Hallam, Arthur. 1862. In *Remains in Verse and Prose of Arthur Henry Hallam*, ed. Henry Hallam. London: John Murray.
Horne, Richard Henry. 1833. *Exposition of the False Medium and Barriers Excluding Men of Genius From the Public*. London: Effingham Wilson.
———, ed. 1844. *A New Spirit of the Age*. 2 vols. London: Smith, Elder, and Co.
Kelley, Phillip, et al., eds. 1984–2016. *The Brownings' Correspondence*. 23 vols. Winfield: Wedgestone.
Landon, Letitia. 1839. *Poetical Works*. Philadelphia: Carey and Hart.
London Dialectical Society. 1871. *Report on Spiritualism of the Committee of the London Dialectical Society*. London: Longmans, Green, Reader, and Dyer.
Macaulay, Thomas Babington. 1843. *Critical and Historical Essays Contributed to the Edinburgh Review*. 3 vols, 2nd ed. London: Longman, Brown, Green, and Longmans.
Martineau, Harriet. 1845. *Letters on Mesmerism*. 2nd ed. London: Edward Moxon.
Mill, John Stuart. 1981. In *Collected Works*, ed. John Robson. Toronto: University of Toronto Press.
Sandby, George. 1844. *Mesmerism and Its Opponents*. London: Longman, Brown, Green, and Longmans.
Tennyson, Alfred. 1987. *The Poems of Tennyson*, ed. Christopher Ricks, 3 vols, 2nd ed. Harlow: Longman.
Townshend, Chauncey Hare. 1821. *Poems*. London: Thomas Boys.
———. 1827. *The Reigning Vice: A Satirical Essay in Four Books*. London: Longman, Rees, Orme, Brown, and Green.
———. 1829. An Essay on the Theory and the Writings of Wordsworth. *Blackwood's Magazine*, vol. 26, 453–63, 593–609, 774–88, 894–910. Edinburgh: Blackwood.
———. 1840. *Facts in Mesmerism, with Reasons for a Dispassionate Inquiry into It*. London: Longman, Orme, Brown, Green, and Longmans.
———. 1851. *Sermons in Sonnets*. London: Chapman and Hall.
———. 1854. *Mesmerism Proved True, and the Quarterly Reviewer Reviewed*. London: Thomas Bosworth.
Wordsworth, William. 1985. In *The Fourteen-Book Prelude*, ed. W.J.B. Owen. Ithaca: Cornell University Press.

Secondary Sources

Auerbach, Nina. 1982. *Woman and the Demon: The Life of a Victorian Myth*. Cambridge, MA: Harvard University Press.

Barrow, Logie. 1986. *Independent Spirits: Spiritualism and English Plebeians, 1850–1910*. London: Routledge and Kegan Paul.

Bate, Jonathan. 2004. *John Clare: A Biography*. London: Picador.

Blake, Kathleen. 1986. Elizabeth Barrett Browning and Wordsworth: The Romantic Poet as a Woman. *Victorian Poetry* 24 (4): 387–398.

Chapman, Alison. 2003. "A Poet Never Sees a Ghost": Photography and Trance in Tennyson's Enoch Arden and Julia Margaret Cameron's Photography. *Victorian Poetry* 41 (1): 47–71.

Colley, Ann. 1983. *Tennyson and Madness*. Athens, GA: University of Georgia Press.

Connor, Steven. 2010. All I Believed Is True: Dickens under the Influence. *19: Interdisciplinary Studies in the Long Nineteenth Century* 10. https://doi.org/10.16995/ntn.530.

Dally, Peter. 1989. *Elizabeth Barrett Browning: A Psychological Portrait*. London: Macmillan.

Felluga, Dino. 2005. *The Perversity of Poetry: Romantic Ideology and the Popular Male Poet of Genius*. New York: New York State University Press.

Fulford, Tim. 2004. Conducting the Vital Fluid: The Politics and Poetics of Mesmerism in the 1790s. *Studies in Romanticism* 43 (1): 57–78.

Handley, Sasha. 2012. Sleepwalking, Subjectivity and the Nervous Body in Eighteenth-Century Britain. *Eighteenth-Century Studies* 35 (3): 302–323.

Hughes, William. 2015. *That Devil's Trick: Hypnotism and the Victorian Popular Imagination*. Manchester: Manchester University Press.

Kaplan, Fred. 1975. *Dickens and Mesmerism: The Hidden Springs of Fiction*. Princeton: Princeton University Press.

Karlin, Daniel. 1989. Browning, Elizabeth Barrett, and 'Mesmerism'. *Victorian Poetry* 27 (3/4): 65–77.

Kurshan, Ilana. 2006. Mind Reading: Literature in the Discourse of Early Victorian Phrenology and Medicine. In *Victorian Literary Mesmerism*, ed. Martin Willis and Catherine Wynne, 17–38. Amsterdam: Rodopi.

Lamont, Peter. 2004. Spiritualism and a Mid-Victorian Crisis of Evidence. *The Historical Journal* 47 (4): 897–920.

Matthew, H.C.G., and Brian Harrison, eds. 2004. *Oxford Dictionary of National Biography*. 60 vols. Oxford: Oxford University Press.

Mitchell, Leslie. 2003. *Bulwer Lytton: The Rise and Fall of a Victorian Man of Letters*. London: Hambledon and London.

Morlier, Margaret. 1990. The Death of Pan: Elizabeth Barrett Browning and the Romantic Ego. *Browning Institute Studies* 18: 131–155.

Owen, Alex. 2004. *The Darkened Room: Women, Power, and Spiritualism in Late Victorian England*. Chicago: University of Chicago Press.

Paroissien, David. 1971. Mrs. Browning's Influence on and Contribution to A New Spirit of the Age (1844). *English Language Notes* 8: 274–281.

Ruston, Sharon. 2013. Creating Romanticism: Case Studies in the Literature. In *Science and Medicine of the 1790s*. Basingstoke: Palgrave.

Ruth, Jennifer. 1999. "Gross Humbug" or "the Language of Truth"? The Case of the Zoist. *Victorian Periodicals Review* 32: 299–323.

Sage, Victor. 1988. *Horror Fiction in the Protestant Tradition*. New York: St Martin's Press.

Sanders, Charles. 1942. *Coleridge and the Broad Church Movement*. New York: Octagon.

Shuttleworth, Sally. 1996. *Charlotte Brontë and Victorian Psychiatry*. Cambridge: Cambridge University Press.

Thomas, Donald. 1982. *Robert Browning: A Life Within Life*. New York: Viking.

Winter, Alison. 1998. *Mesmerized: Powers of Mind in Victorian Britain*. Chicago: University of Chicago Press.

CHAPTER 6

'The Madness': Inspiration and Insanity in Spasmodic Poetry, 1851–1855

Drawing upon the works of Rudy, Blair, and Tucker, this chapter discusses the rise of 'spasmodic' poetry in the 1850s, and argues that the insistence of the spasmodic poets that poetic inspiration transcended all other forms of authority can be understood as a reaction to the medicalisation and pathologisation of poetic talent in British culture over the course of the previous two decades. It explores the depiction of poetic genius in the writings of 'spasmodic' writers such as Horne, Dobell, Bailey, and Bigg, and contrasts their views with those of the *Blackwood's Magazine* writers who critiqued them: of these, special attention is given to the writings of D.M. Moir, an influential poet and critic who was also a practising doctor. Finally, it considers the works published by Browning and Tennyson during the 1850s as responses to the spasmodic movement, and as the culminations of their respective trajectories away from the models of poetic inspiration which had so attracted them both a quarter-century before.

'Scorned and Visionary Men': Richard Horne and the 'Spasmodic' Cult of Poetic Genius

The so-called 'spasmodic' poets, who were briefly but intensely popular during the early 1850s, have long been something of an embarrassment to historians of Victorian literature. The movement's politics were radical and often pro-Chartist, and its signature traits were its emphasis on

© The Author(s) 2019

J. Crawford, *Inspiration and Insanity in British Poetry*,
Palgrave Studies in Literature, Science and Medicine,
https://doi.org/10.1007/978-3-030-21671-9_6

subjective emotional experience, especially as mediated through the body, and the grand claims it made for the power and importance of poets. As Rudy has shown, the spasmodics hoped to use the rhythms of their poetry to physically reshape the minds of their readers for the better: as he puts it, 'In a profoundly democratizing gesture, poetry in the spasmodic model seems no longer limited to an elite few but is directed instead to the human body and universal experiences' (Rudy 2009, 79–80; Boos 2004, 553). Philip James Bailey's *Festus* (1839), which served in many respects as the foundational text of the spasmodic movement, was highly praised in its day, and Tennyson seems to have honestly believed that Bailey was a greater poet than himself (Tennyson 1982, 1:265, 1:283). And yet in the mid-1850s the movement withered under the impact of critical attacks from *Blackwood's Magazine*, and its fortunes have never since been revived.

In this chapter, I argue that spasmodism represented an attempt to reconcile the competing discourses of medical psychology and Romantic genius, by asserting simultaneously that poetic genius was defined by physiological abnormality and vulnerability to illness, *and* that it nonetheless possessed transcendent authority and power. In doing so, however, the spasmodic poets left themselves vulnerable to being dismissed as merely sick or mad by their critics—a tactic whose success is visible in the very fact that they are still known today as 'spasmodics', a name bestowed upon them in mockery by their enemies. The poetry of the spasmodics rested upon the assumption that true poetic genius would be self-evident, and that the brilliance of their works would prove to their readers that they were inspired rather than insane. But as soon as faith in their inspiration wavered, their claims of prophetic authority proved all too easy to reinterpret as mere delusions, especially as the spasmodics themselves placed so much emphasis upon the abnormal condition of the minds and bodies of poets. Indeed, it was for this reason that their credibility declined so quickly after 1854.

Spasmodism was, in some respects, an outgrowth of trends within British literary culture that had been ongoing since the 1830s. Its fascination with the figure of the poet as inspired visionary resulted from the growing influence of Wordsworth and Shelley: as Tucker puts it, 'spasmodism was Romanticism in bells and whistles' (Tucker 2004, 429). Its preoccupation with intense subjectivity and poetic psychology owed something to the writings of Fox and Mill, while its fixation upon the idea of poets as world-changing heroes was largely derived from Carlyle. The continuity between the interpretations of poetic genius articulated by the

Fox circle in the 1830s and those expressed among the spasmodics in the 1850s is demonstrated by the fact that Richard Horne figured prominently in both without making any material changes to his views, as the same opinions that he expressed in his 1833 *Exposition* served him just as well as a spasmodic writer 20 years later.

Horne had attained a measure of poetic fame after the publication of his epic, *Orion*, in 1843. In line with his theory that the British publishing industry was setting up false barriers between writers of genius (such as himself) and the reading public, he sold the poem for the loss-making price of one farthing per copy, thus guaranteeing himself an extensive sale (Matthew and Harrison 2004, 28:142). In 1851 he followed this up with a novel, *The Dreamer and the Worker*, which demonstrated that his faith in the importance of literary genius had not wavered since the 1830s. *The Dreamer and the Worker* is a typically 'spasmodic' work insofar as it combined a commitment to democratic politics with a deeply elitist belief that visionary geniuses are the true and secret rulers of the world, because they originate the grand ideas which the rest of the population inevitably proceeds to act out, being unable to develop any such ideas themselves—a view essentially identical to that of the conservative Bulwer-Lytton, who held that artists constituted the world's true aristocracy (Mitchell 2003, 132–3, 174). As Tucker notes, the spasmodics opposed political tyranny, but 'the terms of spasmodism furnished no other remedy for official imposture than the fantasized imposition of a different single vision, namely the poet's' (Tucker 2004, 440).

In the preface to his novel, Horne described it as a work concerned with 'the social and political position and progress of the working classes', aligning it with the then-popular genre of the 'social problem novel' exemplified by the works of Gaskell and Disraeli. Unlike them, however, Horne maintained that the solutions to the problems of the age were to be found not in politics or economics, but in poetry: and, in his preface, he explained that what set his novel apart from theirs was that 'it never fails, on all fair opportunities, to claim for the literary man, and public teacher, a due recognition of his order' (Horne 1851, 1:iii–iv). He argued that although the life of a 'visionary and a dreamer' might appear to be useless, such people were in fact at least as important to the world as workers, because it was their dreams that made possible the future work of others (Horne 1851, 1:vi). Sure enough, the novel's two heroes—the worker, Harding, and the Shelley-quoting dreamer, Archer—swiftly prove themselves to be complementary human types, and both are shown to be vastly superior to 'the great masses of mankind, or average human character' (Horne 1851, 1:11). The novel's title might be taken to imply an equality of value between these two characters, but in practice it is heavily dominated by 'the dreamer', Archer, who shapes and directs Harding's mental development throughout the narrative. The value of Harding's practical skills is obvious, but the novel repeatedly points out that the dreams of Archer are equally crucial—and, indeed, large sections of the novel are given over to scenes in which Archer lectures the other characters about the true value of dreams, music, history, and poetry, and of the importance of keeping an open mind about mesmerism, while they listen in gratitude and awe (Horne 1851, 1:93–107, 1:218–41, 1:249–80, 1:302–6, 2:42–6, 2:90–5, 2:158–60).

As Archer himself proclaims:

> The practicalities of the world are excellent, if applied to good purposes; but the good purposes are derived from the scorned and visionary men—the recluse theorist—the unappreciated dreamer whose work is all in the shade. (Horne 1851, 1:96–7)

It is a peculiarly specific division of labour that Horne insists upon. What is involved is not anything as simple as a division between physical and mental labour, for Horne makes clear that he considers scientists, administrators, and their ilk as merely so many 'workers', concerned with 'the practicalities of the world' (Horne 1851, 1:97–8). Instead, the labours of all such practical individuals are collectively set against the works of 'unappreciated dreamer[s]' who, *simply by the act of dreaming*, achieve something as important as all of them put together. When another character suggests to Archer that 'the devotee of beautiful abstractions is very likely to rest contented with the abstraction', he replies that this makes no difference:

> Suppose he did—what then? What can limit the *influence* upon others? He is only the originator of an abstraction—that is *his* part—his work in the world—why should he not rest in it? (Horne 1851, 1:98)

In other words, the whole duty of a 'visionary' is simply to dream up a 'beautiful abstraction' and then to communicate it to a few other people (as the interminably monologuing Archer is so fond of doing—especially to Harding, who 'listen[s] with profound silence and attention, seeming to devour every word') (Horne 1851, 1:77). Having done this, he is free to 'rest' in the knowledge that he has accomplished something quite as valuable as any quantity of physical or mental labour. Archer's genius is not immediately appreciated by society, and his 'Philosophical Novel' is rejected by the publishers, just as Horne's *Exposition* would have predicted; but, none the less, the novel makes clear that the '*influence*' of his dreams upon humbly receptive men such as Harding will ultimately change the world for the better (Horne 1851, 1:172–3, 2: 261–4).

As I have discussed, Horne had been making claims like this for decades; but by the 1850s many other writers had developed equally exalted ideas about the power and status of visionary dreamers and poets. This was largely due to the growing popularity of Carlyle, whom Barrett referred to in 1845 as 'the great teacher of the age' (Barrett Browning and Browning 1990, 26). As early as *Sartor Resartus*, Carlyle had asserted that 'the Poet and inspired Maker' was 'a Hierarch, therefore, and Pontiff of the World' (Carlyle 1975, 169). In his 1840 lectures on *Heroes and Hero-Worship* he had argued that all kinds of 'Great Men'—kings, priests, prophets, and poets—were simply different manifestations of the same heroic principle, and that great poets could just as easily have been great kings or prophets if they had been born into different historical circumstances:

> I confess, I have no notion of a truly great man that could not be *all* sorts of men. The Poet who could merely sit on a chair, and compose stanzas, would never make a stanza worth much. He could not sing the Heroic warrior, unless he himself were at least a Heroic warrior too. I fancy there is in him the Politician, the Thinker, Legislator, Philosopher:—in one or the other degree, he could have been, he is all these. (Carlyle 1866, 71)

This faith in the world-changing power of poets can be seen in the works of many other mid-century writers, for whom 'poet' and 'prophet' increasingly seemed to act as interchangeable terms. Bulwer-Lytton claimed in 1841 that law inevitably followed literature rather than vice versa, as 'the Lawgiver is compelled to redress what the Poet has lifted into esteem' (Mitchell 2003, 174). In *The Bards of the Bible* (1851), George Gilfillan, who would go on to play a crucial role in popularising the works of the spasmodics, argued that the Bible could be understood as a work of poetry: 'not as a revelation of *special*, but as a poem embodying *general* truth' (Gilfillan 1851, xix). For Gilfillian, it thus followed that the prophets were to be regarded primarily as poets—a change which he did not regard as reducing their status in any way. January Searle, in his 1852 *Memoirs of William Wordsworth*, rapturously described Wordsworth as an 'oracle of the Infinite', 'the singer of a new gospel of life to the world' who had rescued 'the old Harp of the Bards' from the 'uninspired', and compared him to Moses after he had seen God, complete with a 'prophet's mantle'. Sounding very much like Horne and Carlyle, he went on to assert that when confronted with a true poetic genius, 'we are to accept his law, not he ours':

> For of all men the poet is highest and noblest. He is the awful Seer, who unveils the spirit of Nature and looks with solemn and unscathed eyes upon her naked loveliness and terror. The Beloved of God, he is admitted into the presence of the Invisible, and reports, in such wild and strange words as he can find, the sights he has beheld there. He is the renovator of man and Nature. He lifts the human soul upon his daring wings, and carries it into light and immortality. In his words we behold the re-creation of the universe. […] He is master of all law, and no critic can trammel him. (Searle 1852, 13–14, 18, 24–6)

Similar views were articulated by many contemporary poets. Philip James Bailey, whom Horne and Barrett had praised in *A New Spirit of the Age*, had first become famous for the success of *Festus* (1839), and then infamous for the failure of its sequels *The Angel World* (1850) and *The Mystic* (1855): in all three works he wrote repeatedly of the dignity of inspired bards such as himself, who brought spiritual truth into the world through their poetry. In much the same vein, Gilfillian's protégé Alexander Smith's *A Life-Drama* (1853) described the struggles of a would-be poet to become a prophetic 'poet-king' who might 'sun-crown the age', while John Stanyan Bigg's 'dramatic poem', *Night and the Soul* (1854), depicted yet another poet struggling with despair, and eventually triumphing over it, in order to pursue his destiny as a true 'poet-king' whose poetry would help to reform the world (Smith 1859, 24; Bigg 1854, 170).

All of these poets shared Horne's belief in the special dignity of poetic genius. Bailey wrote in *Festus* of how 'Poets are henceforth the world's teachers'. Smith wrote grandly of the 'mighty Poet whom this age shall choose / To be its spokesman for all coming times', while Bigg described 'the mighty bards' in rapturous terms, claiming that 'the great universe streams through their souls / Like a wide river of perpetual light' (Bailey 1845, 287; Smith 1859, 24; Bigg 1854, 130). One long section of *Festus* is given over to a scene in which the titular protagonist lectures a nameless 'student' about the true nature of poetic genius, which he describes in increasingly prophetic terms:

> And he who means to be a great bard, must
> Measure himself against pure mind, and fling
> His soul into a stream of thought, as will
> A swimmer hurl himself into the water.
> But never swimmer on the stream, nor bird
> On wind, feels half so strong, or swift, or glad,
> As bard borne high on his mind above himself;
> As though he should begin a lay like this,
> Where spiritual element is all;
> Through chafing thought, as bough bough, till all burn,
> Like the star-written prophecies of Heaven. (Bailey 1845, 285)

There is a telling shift here, as the need for the 'bard' to 'Measure himself against *pure* mind' gives way to praise for his ability to be 'bourne high on *his* mind above himself'. (My italics.) It is only *his own* mind that the 'bard' needs to measure himself against, because—as Bailey explains—the true bard makes his own rules, and in any case he will never be understood by ordinary people (Bailey 1845, 279–81). Horne had claimed that only one genius could recognise another, but Bailey goes one step further, and argues that the only person who can properly evaluate the bard's inspiration is *himself*:

> Believe thou art inspired, and thou art.
> Look at the bard and others; never heed
> The petty hints of envy. If a fault
> It be in bard to deem himself inspired,
> 'Tis one which hath had many followers
> Before him. (Bailey 1845, 287)

Festus thus becomes the poetic equivalent of Browning's Johannes Agricola, asserting that the simple fact that he believes himself to be inspired is, itself, proof of the validity of his inspiration, just as Johannes Agricola maintains that his faith in his own salvation proves that he is actually saved. This solipsistic self-referentiality is actually enacted within the text of *Festus*, as its protagonist delivers a series of lengthy speeches about the life, works, and genius of a great and much-neglected poet, only to ultimately reveal that *he himself* is the unknown poet whose career he has been describing in such glowing terms (Bailey 1845, 289). Viewed by his own narcissistic standards, there is actually a great deal of logic to this. After all, who except Festus himself could possibly be capable of recognising and understanding the true grandeur of his own poetic genius?

For Carlyle, the poet had been only one kind of 'Great Man' among many, to be ranked alongside great generals, prophets, and kings. In the writings of Bailey and his imitators, however, poetic genius was increasingly represented as subsuming and transcending all other forms of greatness, to the point where the difference between writing about great deeds and actually performing them threatened to become entirely lost. The protagonist of Smith's *Life-Drama*, Walter, even declares that the act of writing a poem which adequately articulated the 'great complaint' of the age would, *in and of itself*, suffice to end the very 'complaint' that it described:

> I do believe
> When it is fully sung,—its great complaint,
> Its hope, its yearning, told to earth and heaven,—
> Our troubled age shall pass, as doth a day
> That leaves the west all crimson with the promise
> Of a diviner morrow[.] (Smith 1859, 85–6)

Sydney Dobell's *Balder* (1854) took the same theme even further, with its depiction of a poet who decides to commit murder so that he can write a poem about death—a poem which he believes will be of such immense benefit to humanity as to be well worth the sacrifice of a human life. As Tate notes, 'Balder's overweening conception of the "royal" power of his poet's psyche is so heightened here that he implies that his sense experience, and the outside world, can be reshaped in order to conform to his imaginings' (Tate 2012, 82).

The insistence of these writers on the omnipotence of poetic genius resulted from a fusion of Carlyle's assertion that a truly great poet must also embody all other forms of human greatness with Shelley's claim that poets were 'the unacknowledged legislators of the World' (Carlyle 1866, 71; Shelley 1977, 508). But it must also be understood within the context of the period's growing scepticism about the nature and validity of poetic inspiration. As I have discussed, the kind of 'visionary' imagination which these writers praised so highly had been increasingly medicalised and pathologised during the 1830s and 1840s. These writers responded to the question of whether poetic inspiration was anything more than a symptom of mental disease, firstly, by insisting that inspired poets spoke with greater authority than anyone else, doctors very much included, and, secondly, by staging exaggerated displays of 'inspired' poetic writing in order to demonstrate that they themselves possessed the kind of prophetic power and authority that their works described. As late as 1853, this tactic seemed to be working, with Smith and Bailey, in particular, widely hailed as two of the greatest poets of the age. Over the next few years, however, confidence in their pseudo-prophetic status suffered an abrupt collapse, the speed of which suggests just how brittle such claims of inspired poetic authority had become (Tate 2012, 90). Tellingly, when their critics set out to discredit them, they drew heavily on the discourse of medicine: and soon these would-be poet-kings had come to be known merely as 'spasmodics', their supposedly inspired works dismissed as spasms of their diseased brains (Aytoun 1854, v; Faas 1988, 138–9).

'Egotistic, Self-Contained, and Sophistical': The *Balder* Controversy

As Tate has pointed out, the spasmodics were not opposed to the idea that poetic genius had a physiological basis, rooted in the blood, brain, and nerves, any more than Mill, Horne, and Hallam had been two decades before (Tate 2012, 79–80). Furthermore, as Blair notes, they themselves often described poetic inspiration in terms of sickness or madness, introducing what she calls 'a new kind of poetic hero, who is indeed nervous, emotional, and prone to spasmodic attacks' (Blair 2004, 473–4). It was this notion that George Eliot later drew upon in 'The Lifted Veil', published in 1859 but set in 1850, whose rather spasmodic protagonist, Latimer, combines physical and mental infirmity with supernatural powers of perception. Explaining that 'I thoroughly disliked my own *physique*, and nothing but the belief that it was a condition of poetic genius would have reconciled me to it', Latimer clings to the fact that 'in genuine biographies I had read of the subtilising or exalting influence of some diseases on the mental powers', but continues to worry that his apparent visions might 'rather be a disease—a sort of intermittent delirium, concentrating my energy of brain into moments of unhealthy activity, and leaving my saner hours all the more barren' (Eliot 1999, 10–14).

Committed to a physiological model of poetry in which the greatest poets would be the most feverishly passionate of men rather than the calmest, the spasmodics acknowledged that poetic genius carried with it the risk of madness: indeed, the struggle to achieve artistic greatness without going mad was essentially the grand theme of most spasmodic poetry. *Festus*, *The Mystic*, *A Life-Drama*, *Balder*, and *Night and the Soul* were all studies in the tribulations of genius, depicting the mental ordeals which truly superior individuals had to undergo on their journey towards artistic and spiritual mastery. Like Horne before them, Smith, Bigg, and Dobell saw these ordeals as being so trying that poets might often veer dangerously close to insanity: there is a great deal of fever and sickness in spasmodic poetry, and spasmodic poet-heroes usually suffer enormous amounts of physical and mental distress. As Smith later remarked, the 'peculiar avocation' of men of letters made them uniquely ill-suited to dealing with life's difficulties and hardships, and 'makes sad work with the fine tissues of the nerves' (Smith 1863, 149–50, 157).

The spasmodics, then, had thoroughly absorbed the medicalised accounts of poetic genius common to the period, and this rendered their attempts to claim inspired prophetic authority for themselves and their protagonists highly vulnerable to being dismissed as symptoms of what Dobell's Balder calls 'The populous fever of a poet's brain' (Dobell 1854, 168). Indeed, the controversy which erupted after the publication of *Balder* demonstrates just how easily spasmodic inspiration could be mistaken for diseased insanity, and vice versa. Like Festus, the eponymous Balder is convinced that he is a great poet (complete with a 'prophetic heart'), whose destiny is to be 'the King of men, and on the inform / And perishable substance of the Time / Beget a better world'. As such, he believes that it is his duty to write a great epic poem, which will apparently have the power to lead humanity into this glorious future (Dobell 1854, 9–18). But in order to do this, he believes that it is necessary for him to gain first-hand experience of death, on the grounds that true poetry should be a record of authentic subjective experience. He thus decides to murder his infant daughter, viewing her death as an acceptable price to pay for the benefit that the poem which results from it will bring (Dobell 1854, 63–71; Morton 2008, 850–1). It is very unclear whether he actually *does* murder the child; but she certainly dies, either by illness or by his hand, leading Balder's wife to decline into madness as a result of her grief and his neglect. Complaining that her mental distress is preventing him from writing his great poem, and thus holding back his leadership of 'the bannered host of man', Balder decides to kill her, too (Dobell 1854, 230–1, 248–71). And yet despite his wild claims about the incredible benefits that the world will reap from his never-completed poem, *Balder* never offers any sign that the poems of its protagonist actually have the extraordinary powers which he attributes to them. Like Festus, Balder apparently thinks that his belief in his own poetic greatness is the only proof he needs—and that this belief effectively gives him a licence to kill.

Balder was heavily attacked by contemporary critics, who viewed its story as being immoral. Stung by these criticisms, Dobell responded in the preface to his second edition that he did not 'personally admire' Balder, and that his immorality was precisely the point of the poem:

> I understand that the public press have described my hero to be egotistic, self-contained, and sophistical, imperfect in morality, and destitute of recognised religion, mistaken in his estimate of his own powers and productions, and sacrificing to visionary hopes and dreamy distant philanthropies the blessing that lay in his embrace, and "the duty which was nearest".
>
> This is precisely the impression which I wished the readers of this volume to receive, and I owe some acknowledgement for such loud and emphatic testimony that exactly what I desired to attain has been attained. (Dobell 1854, iv–v)

Balder, Dobell explained, was only the first part of a longer poem, which was to lead its hero 'from Doubt to Faith, from Chaos to Order', and its protagonist was not 'a model' but 'a warning' (Dobell 1854, iv–v). Dobell's modern critics, such as Rudy, Boos, and Blair, have affirmed that the morally serious Dobell would never have deliberately written a poem in praise of murder, and have argued that *Balder* should be understood in the same way as 'Porphyria's Lover', as what Boos calls 'a post-romantic critique of its Faustian/Byronic protagonist which would point an implicit moral of the dangers of overweening self-regard' (Rudy 2009, 87–91; Boos 2004, 569; Blair 2004, 486). Yet the strongly negative reception of the poem indicates that many of Dobell's readers clearly read it as endorsing Balder's murderous narcissism rather than critiquing it, and Aytoun subsequently attacked it as a paradigmatic example of spasmodic amorality.

Dobell's problem was that *Balder* sent extremely mixed messages to its audience. Balder rants, rages, treats women appallingly, experiences worrying physical symptoms, and makes insane-sounding claims for the world-changing power of his poetry—but *all* spasmodic poet-heroes did those things, and in works such as *Festus* and *A Life-Drama* those very traits were represented as signs of their authentic poetic genius rather than as proofs of their insanity. Balder may be in a state of mental 'Doubt' and 'Chaos', but the claims which he makes for the power of his poetry to 'Beget a better world' and make him 'The King of men' were probably not intended by Dobell to be read as symptoms of his mental disturbance—after all, as I have discussed, Smith, Bigg, Bailey, and Horne had all made very similar claims for their respective poet-heroes. Dobell himself directed 'the really earnest enquirer' to 'the elements of my hero as they exist, uncombined or undeveloped, in the much-observed and well-recorded characters of men who have been, and have more or less deserved to be, praised, loved, followed, and revered', suggesting sources such as 'the Autobiography of Haydon' and 'the letters of Keats', and thus implying that he viewed Balder as a flawed but authentic genius rather than a straightforwardly delusional figure (Dobell 1854, vi–vii). Even so acute a reader as Charlotte Brontë was left uncertain as to whether or not Balder represented Dobell's 'cherished ideal of the true great poet' (Boos 2004, 570).

Given that the mental traits associated with Balder's poetic talents were the same traits that were associated with his murderous delusions—both talent and delusion being understood to result from the 'fever' of his 'poet's brain'—it is understandable that Dobell's readers should have experienced some confusion over which parts of Balder's delirium they were supposed to admire as proof of the quality of his genius, and which they were supposed to deplore as evidence of his unhealthy and imbalanced mind. Horne, Bailey, Smith, and Bigg had established such a powerful connection between poetic inspiration and mental disturbance that by the time *Balder* appeared it was by no means obvious that its protagonist's bizarre behaviour, infanticide and all, was meant as anything other than a completely sincere depiction of the mental agonies associated with true poetic genius. It was this apparent interchangeability between inspired 'spasmodic' poets and criminal lunatics which, over the next few years, was seized upon by the critics of spasmodism, with results that proved disastrous for the movement as a whole.

'To Hell, Ye Empirics!': Medicine and Spasmodic Poetry

Despite accepting that poetic genius had a physiological basis in the heart and nerves, the spasmodics strongly resisted the idea that this meant poetic inspiration could be better understood by doctors than by poets. In *Night and the Soul*, when the poet Alexis, in the throes of despair, describes to his friend Ferdinand his vision of a meaningless universe—'Last night I dreamed the universe was mad'—Ferdinand's reductive reply of 'Nightmare, my friend—and heavy suppers—hey?' highlights the inadequacy of such materialistic explanations to account for the visions of the poet (Bigg 1854, 132, 135). Alexis' imagination must heal itself, just as Walter's does in *A Life-Drama*, without the help of medicine: it is only poetic restoration of faith, and not a mere change of diet, that can save him from declining into insanity. The authority of doctors over poets is even more comprehensively denied in *Balder*, in which the hero declares that the condition of his soul transcends all possible medicine:

> [W]hat eye
> Infalliable shall find, what lip shall name
> My hid disease? To hell, ye empirics!
> And burn your statutes! (Dobell 1854, 215)

The frequently expressed hope of these writers was that their poetry would have the power to reform and redeem the age, rescuing it from its materialistic stupor, and forcing it to recognise that the prophetic authority of true poetic genius transcended that of any scientist or doctor. However, as Armstrong notes, their poetry was routinely described by unsympathetic critics as being the product of madness: '"Rant", "Bedlam", "epilepsy", "lunatic" were common epithets' (Armstrong 1996, 253). As an 1848 article in *The Times* sternly warned, 'Even the madness of our poets must have its method, or be dismissed to the asylum' (Matthews 1971, 321).

A clear example of this line of criticism can be seen in the lectures which D.M. Moir—better known to the British public as 'Delta', the name under which he wrote for *Blackwood's Magazine*—delivered at the Edinburgh Philosophical Institution in 1850–1851 (Coyer 2016, chap. 3). As a poet and critic who was also a practising doctor, Moir had firm views on what kinds of poetry were and were not compatible with good mental health. All truly great poems, he argued, could be understood by ordinary people, and judged by what he called 'a jury of common sense men'. 'You may rely upon the test', he wrote: and when poets moved away from this standard, becoming obscure and 'visionary' in pursuit of some private subjective insight, both their poetry and their mental health was likely to suffer as a result (Moir 1852, 231).

As his appeal to non-elite audiences suggests, Moir was no enemy to most of what is now called 'Romanticism'. He praised Burns, Wordsworth, Coleridge, and Scott as the saviours of British poetry, delivered mixed but ultimately positive verdicts on Keats and Byron, and thought that Tennyson thoroughly deserved to be poet laureate. His views, however, placed him on a collision course with the spasmodics: and, over the course of his lectures, Moir narrated what he saw as being the story of British poetry's rise and fall. In the first years of the nineteenth century, he asserted, 'poetical literature' had been rescued from the 'artificial' style of Hayley and Darwin by the works of Burns, Wordsworth, and Scott; but since then it had lost its way, seduced from its true path by German metaphysics, leading him to conclude sorrowfully that 'poetry is at present prostrated' (Moir 1852, 11, 326, 335). The chief villain in Moir's story was Shelley, whom he described in terms which drew more upon his medical training than upon his knowledge of literature:

> It is needless to disguise the fact, and it accounts for all—his mind was diseased: he never knew, even from boyhood, what it was to breathe the atmosphere of healthy life, to have the *mens sana in copore sano*. His sensibilities were over-acute; his morality was thoroughly morbid; his metaphysical speculations illogical, incongruous, incomprehensible—alike baseless and objectless. (Moir 1852, 231–2)

'It accounts for all': medicine, here, is presented as definitively trumping poetry when it comes to explaining Shelley's poetic career. Shelley, Moir claimed, had perceived life as 'a phantasmagorial pictured vision', and 'seemed to take an insane delight in selecting, for poetical illustration, subjects utterly loathsome and repulsive'. As a result, his 'mystical speculations' were 'as little worth the attention of reasoning and responsible man as the heterogeneous reveries of nightmare' (Moir 1852, 232). And yet despite being diseased, unreasonable, morbid, illogical, unhealthy, and insane, 'a wayward, misguided, and singular man' whose genius had been 'of a peculiar and unhealthy kind', Shelley had exercised an enormous influence over subsequent poets, with the result that they had come to 'flatter themselves with the belief that their reveries are realities'—an accusation which, as I have discussed, could certainly be levelled with some justice against the poet-protagonists of many spasmodic poems. As Moir complained, 'Simple utterance of feeling—with a mystical commentary on such utterance—is all that the purest disciples of this newest of our schools aspire to' (Moir 1852, 225–6, 231, 331, 333).

For Moir, Shelley's influence constituted nothing less than a diseased strain within British poetry, a strain which he saw as having become increasingly prominent over the course of the 1840s. He attributed its success to a poetic revolt against the increasingly scientific character of the age, a revolt which he saw as entirely misguided. In his view, science and poetry were not rival forces battling for mastery over the human soul: instead, they were equally valuable but entirely separate pursuits, to be divided from one another and cultivated separately, just as he divided his own time between his medical practice and his poetic career (Moir 1852, 323, 331). But the poets of the day had failed to understand this, and in their revolt against scientific rationalism they had consequently fallen into the kind of 'mystical' metaphysics which, in Moir's view, had fatally undermined both Shelley's poetry and his sanity. He thus repeatedly discussed this drift towards metaphysical Shelleyan poetry in terms of madness and disease, complaining, for example, that Barrett Browning's early works had been 'lucid', but that she had since declined into 'morbid excitement' (Moir 1852, 280, 283).

As can be seen from his rather helpless description of it as 'this newest of our schools', Moir did not yet have a name for the kind of writing he was attacking. Part of his difficulty was due to the sheer breadth of his assault, which took in everyone from Carlyle and Emerson to Browning and Barrett Browning: as far as Moir was concerned, anyone whose works were metaphysical and/or obscure was part of the same Shelleyan contagion (Moir 1852, 280, 324–6). He did, however, recognise that the sickness had different degrees: and the 'purest disciple' of all, whom Moir saw as being to Tennyson and Barrett Browning what the 'diseased' Shelley had been to the 'visionary' but still basically sane and healthy Keats, was Philip James Bailey (Moir 1852, 223, 333). Moir's death in July 1851 prevented him from witnessing the full flowering of Bailey's influence in 1853–1855, and it was left to another Blackwood's man, William Aytoun, to describe the resulting school of poetry with the adjective 'spasmodic'. In retrospect, however, the kinds of poetry which most worried Moir in 1851 were clearly forms of spasmodism *avant la lettre*: works in which the judgement of Moir's 'jury of common-sense men' were seen as entirely irrelevant when compared to the personal inspiration of the poet.

What seems to have disturbed Moir about this kind of poetry, with its dedication to 'simple utterance of feeling' and its belief that 'reveries are realities', was its total subjectivity. It did not seek—and claimed not to require—any kind of external validation of its claims, regarding all forms of objectivity as part of the scientific-materialist world-view against which it was in revolt, and leading Moir to paraphrase its core principle as 'A fico for general sympathy and common sense! The man in the moon forever!' (Moir 1852, 331–3). As a result, it was desperately vulnerable to sliding away into lunacy, as Moir's allusion to 'the man in the moon' implies. As Arnold argued in 1854, such poetry 'is *fantastic* and wants *sanity*', and even if such poets were not insane to begin with, they were likely to drift ever further out of touch with reality as long as they persisted with such harmful and misguided poetic practices (Armstrong 1996, 169). Aytoun's parodic drama *Firmillian*, in which a crazed poet commits a whole series of murders because he believes that no one else has any right to pass judgement on the behaviour of a true genius—and because their lives are in any case worthless compared to his poetry, the value of which is, again, something that only he can accurately assess—simply presented the same critique in more exaggerated terms.

When Kingsley satirised the spasmodics in his 1857 novel *Two Years Ago*, it was probably this mutual antagonism between spasmodic poetry and medical science which led him to depict the novel's would-be poet, John Briggs, as a disaffected trainee doctor. Like Moir, Kingsley admired most of the Romantics, including Wordsworth and Byron, but he strongly disapproved of Shelley, whom he saw as having had a disastrous effect upon the development of English poetry (Rutherford 1970, 350, 355; Gill 1998, 61; Ryan 1997, 80). In his 1853 'Thoughts on Shelley and Byron' he had complained, in terms very similar to those used by Moir two years earlier, that the current generation of writers were 'a mesmerizing, table-turning, spirit-rapping, Spiritualizing, Romanizing generation, who read Shelley in secret, and delight in his bad taste, mysticism, extravagance, and vague and pompous sentimentalism', thus connecting the contemporary crazes for mesmerism and spiritualism (with all the occult powers and special insights that they promised to their practitioners) with what he saw as the deeply unhealthy influence of Shelley's poetry (Kingsley 1853, 571).

Like Moir, Kingsley described Shelley's influence over contemporary poetry as a contagious infection, complaining that 'it is Shelley's form of fever, rather than Byron's, which has been of late years the prevailing epidemic'. Shelley, he wrote, had founded 'a spasmodic, vague, extravagant, effeminate, school of poetry', which 'believe[d] that the poet was to regenerate the world, without having settled with what he was to regenerate it'. As a result, his admirers had come to accept what Kingsley saw as the absurd belief that the only thing necessary to redeem the age was to 'Let the poet speak—what he is to say being, of course, a matter of utterly secondary import, provided only that he be a poet; and then the millennium will appear of itself'—a belief which, as I have discussed, was repeatedly articulated by the spasmodic poets themselves (Kingsley 1853, 570–4).

In *Two Years Ago*, Kingsley thus depicted the aspiring poet, Briggs, as an extremely spasmodic character, so convinced of his own special nature that he claims normal life will 'drive me mad'. When asked to apply himself to the work of mixing medicines, he complains:

> It is in vain to ask me to command my temper while I stay here. I am not fit for this work: not fit for the dull country. I am not appreciated, not understood[.] (Kingsley 1906, 21)

Dismissing Briggs as a 'vain, discontented youth', the narrator swiftly diagnoses the cause of his exaggerated sense of self-worth. He has simply been reading too much Shelley:

> Not that the rant was very blameable in a lad of eighteen; for have we not all, while we are going through our course of Shelley, talked very much the same abominable stuff, and thought ourselves the grandest fellows upon earth on account of that very length of ear which was patent to all the world save our precious selves; blinded by our self-conceit, and wondering in wrath why everybody was laughing at us? (Kingsley 1906, 22)

For Kingsley, such Shelleyan enthusiasms are an inevitable part of adolescence; we must go through our 'course' of them, as of any other medicine, and endure the side effects as best we can. Briggs considers himself far too superior a soul for the practice of medicine; but in describing his beloved Shelley as though he was simply one of the drugs which Briggs so resents having to deal with, the narrator, like Moir, deliberately subordinates Shelleyan poetry to medical practice, thus deflating Briggs' attempt to do the opposite. 'Shelley', implicitly, may not be too harmful if taken in moderation; during a 'course' of him users may suffer delusions of being 'the grandest fellows upon earth', but ultimately they will emerge as wiser and more mature individuals. Briggs, however, just doesn't know when to stop: his 'course of Shelley' never ends, and in consequence the very drugs which he cannot be bothered to master end up mastering him. In the later parts of the novel he declines into opium addiction and mental infirmity until 'his nerves and brain are all but ruined', reducing him to nothing more than 'a very curious psychological study' (Kingsley 1906, 396, 410). As Blair notes, his 'abandonment of medicine for poetry symbolically suggests his decline into disease' (Blair 2004, 485).

Blair and Tucker concur in seeing the crucial charge levelled against the spasmodics by their critics as being one of *affectation*. The spasmodic project depended upon its ability to convince readers that its poet-heroes (and their authors) were authentic geniuses, whose oddities and infirmities were inseparable from their poetic talents. Kingsley, Moir, and Aytoun all responded by arguing that the abnormal behaviour of such individuals were mere affectations: thus Tucker's argument that Aytoun's key achievement in *Firmillian* was 'mimicking [spasmodic] techniques of presentation in such detail as to expose them as, precisely, techniques rather than the ingenuous effusions they pretended to be' (Tucker 2004, 440–1). In line with the theories propounded by Conolly, Millingen, Winslow, and other medical psychologists of the era, they furthermore implied that such affectations were likely to terminate in real illness or insanity (Blair 2004, 474, 477, 485). In their account, spasmodism ceased to be a record of the madness that true geniuses must grapple with in order to become great poets, and shrank into a mere tally of the eccentricities that poets affectedly adopted in the hope of being mistaken for geniuses. The spasmodics asserted that the end-point of this process was poetic inspiration, but their critics maintained that it was much more likely to terminate in genuine and unfeigned insanity.

The problem was essentially one of authority. When a poet declared that he was an inspired genius, and a medically informed critic replied that actually he was just a delusional eccentric, whom was the reading public to believe? By 1854 it seemed that the answer was, for the most part, the latter: and the fact that Kingsley's novel was set, as its title declared, *Two Years Ago*, is itself an indication of just how brief the 'spasmodic' vogue ultimately turned out to be. Writing in 1857, Kingsley was able to look back upon it as a wave already past its peak. By the time Thomas Brisbane wrote his biography of Alexander Smith in 1869, he apologised for Smith's 'spasmodic' *Life-Drama* on the grounds that it had been 'the too rapid product of a sensitive mind under the heat of unnatural and unhealthy excitement', and that Smith would never have written such a work if his mind had not been pushed into an 'abnormal state' by getting caught up in radical politics and reading *Festus* (Brisbane 1869, 142–6). If Moir was right to see spasmodism as being a reaction to the scientific character of the age in general, and of mid-century medicine in particular, then it was ultimately not a very successful one. The institutional authority of medical psychology had grown enormously since 1830: and the effectiveness with which Moir, Aytoun, and Kingsley were able to persuade their audience to view the works of these would-be poet-prophets as mere 'spasms' implied that, by the later 1850s, medical diagnosis trumped poetic inspiration after all.

'The Subjective Poet': Browning's 'Essay on Shelley'

As I have discussed, both the allies and the enemies of spasmodism saw it as resulting from the influence of Shelley—but for the poets of the 1850s, Shelley was a much more remote presence than he had been for Tennyson and Browning 20 years before. The sense of Shelley as a heroic figure from a now-vanished age of the world is powerfully communicated by a poem from Browning's 1855 collection, 'Memorabilia':

> Ah, did you once see Shelley plain,
>   And did he stop and speak to you?
> And did you speak to him again?
>   How strange it seems and new!
>
> But you were living before that,
>   And also you are living after;
> And the memory I started at—
>   My starting moves your laughter.
>
> I crossed a moor, with a name of its own
>   And a certain use in the world no doubt,
> Yet a hand's-breadth of it shines alone
>   'Mid the blank miles round about:
>
> For there I picked up on the heather
>   And there I put inside my breast
> A moulted feather, an eagle-feather!
>   Well, I forget the rest. (Browning 1983–2009, 5:257–8)

For the poem's speaker, this momentary encounter with Shelley serves to distinguish the moment in which it takes place from the remainder of the addressee's life, just as the eagle's feather distinguishes the 'hand's-breadth' of moorland upon which it lies from the empty lands around it. The remainder of the addressee's life—the 'living before that' and 'living after'—may have some value on their own terms, just as the moor has 'a certain use in the world no doubt', but in comparison with this moment they are reduced to 'blank miles', empty and meaningless. Shelley is imagined as a creature of quite another order to the men and women around him, and anecdotes of his life, like the fallen feathers of the eagle, are to be valued as proofs of the recent passage of some superior being, far above our heads. (The fallen feather may also echo the story of another prophetic figure, Elijah, who ascended into heaven on a fiery chariot, leaving only his fallen garment lying on the ground as a sign of his ascendance.) But the most telling moment in the poem is the last line, in which the speaker, having found his own poetic symbol—'a moulted feather, an eagle-feather!'—cannot then rise into a poetic ecstasy of his own, as Shelly might have done, imaginatively following the eagle in its flight. Instead, he breaks off, defeated, unable or unwilling to continue, and in place of transcendent vision comes the confession (or possibly the excuse) that 'I forget the rest'. The speaker admires Shelley, but he either cannot or will not follow his path.

Why not, though? As I have discussed, the period was full of poets who were happy to regard themselves as so many Elishas taking up the fallen mantle of Elijah, so why not do so himself? Some insight into Browning's changing views on this subject can be gained from his 1851 essay on Shelley, which he wrote as the preface to what was then believed to be a newly discovered collection of Shelley's letters. In this essay, Browning distinguished between two kinds of poets: the 'objective poet', whose 'endeavour has been to reproduce things external … with an immediate reference, in every case, to the common eye and apprehension of his fellow men' and the 'subjective poet', whose focus was rather different:

> He, gifted like the objective poet with the fuller perception of nature and man, is impelled to embody the thing he perceives, not so much with reference to the many below as to the one above him, the supreme Intelligence which apprehends all things in their absolute truth,—an ultimate view ever aspired to, if but partially attained, by the poet's own soul. Not what man sees, but what God sees—the *Ideas* of Plato, seeds of creation lying burningly on the Divine Hand—it is toward these that he struggles. […] He is rather a seer, accordingly, than a fashioner, and what he produces will be less a work than an effluence. (Browning 1983–2009, 4:424–6)

Shelley, obviously, had been one such 'subjective poet'; Barrett Browning was another, and the spasmodics aspired to be so many more (Haigwood 1986, 99–100). Browning knew that such writers tended to be viewed as the very greatest of poets: as he wrote, 'the subjective [poet] might seem to be the ultimate requirement of every age'. But he insisted that 'objective' poets such as himself, who in place of the visionary talents of the subjective 'seer' possessed a 'faculty of seeing external objects more clearly, widely, and deeply, than is possible to the average mind', were just as important to the history of poetry (Browning 1983–2009, 4:424–6).

In his *Defence of Poetry*, Shelley had described the history of poetry as a constant struggle against the natural tendency of language to degrade into cliché and abstraction. In his account, all true poets were on the same side and had essentially the same duty to fulfil, working together to '[mark] the before unapprehended relations of things and perpetuates their apprehension', and thus prevent language from becoming 'dead to all the nobler purposes of human intercourse' (Shelley 1977, 482). Browning, however, argued that the history of poetry was not as a tug-of-war between poets and the deadening effects of mundanity, but a cyclical progression in which 'subjective' and 'objective' poets had distinct but essential roles to play. The first part of his account conformed to Shelley's, describing how the achievements of each great visionary poet tended to be gradually worn away by their successors over time until almost nothing remained:

> There is a time when the general eye has, so to speak, absorbed its fill of the phenomena around it, whether spiritual or material, and desires rather to learn the exacter significance of what it possesses, than to receive any augmentation of what is possessed. Then is the opportunity for the poet of loftier vision, to lift his fellows, with their half-apprehensions, up to his own sphere, by intensifying the import of details and rounding the universal meaning. The influence of such an achievement will not soon die out. A tribe of successors (Homerides) working more or less in the same spirit, dwell on his discoveries and reinforce his doctrine; till, at unawares, the world is found to be subsisting wholly on the shadow of a reality, on sentiments diluted from passions, on the tradition of a fact, the convention of a moral, the straw of last year's harvest. (Browning 1983–2009, 4:428)

This, for Shelley, would be the moment at which a new 'poet of loftier vision' was called for, to lift these degraded 'successors' back up to their original level. But Browning argued otherwise:

> Then is the imperative call for the appearance of another sort of poet, who shall at once replace this intellectual rumination of food swallowed long ago, by a supply of the fresh and living swathe; getting at new substance by breaking up the assumed wholes into parts of independent and unclassed value, careless of the unknown laws for recombining them (it will be the business of yet another poet to suggest those hereafter), prodigal of objects for men's outer and not inner sight, shaping for their uses a new and different creation from the last, which it displaces by the right of life over death,—to endure until, in the inevitable process, its very sufficiency to itself shall require, at length, an exposition of its affinity to something higher[.] (Browning 1983–2009, 4:428)

The kind of poet described here is, of course, an 'objective' poet such as Browning himself, who reinvigorates the system of poetry not by the power of his 'loftier vision', as Shelley would have prescribed, but by providing his or her successors with 'objects for men's outer *and not inner* sight' (my italics): the raw material that some subsequent 'subjective' poet will one day be able to make use of to express 'its affinity to something higher', thus starting the whole cycle over again. 'Subjective' and 'objective' poets play equal and opposite roles in keeping poetry alive—and a crucial role is thus created for poets who *don't* concern themselves with 'loftier visions', and who dedicate themselves instead to the description of those 'things external' which are perceived not only by would-be prophets and mystics such as the spasmodics, but also by 'the common eye and apprehension of his fellow men'.

Browning's claims for the importance and dignity of 'objective' poetry were, essentially, arguments for the value of poetry which described the external world which ordinary people actually inhabited, rather than just the private visionary insights of the poet. He wrote that 'it would be idle to inquire, of these two kinds of poetic faculty in operation, which is the higher or even rarer endowment', because 'it is with this world, as starting point and basis alike, that we shall always have to concern ourselves':

> [T]he world is not to be learned and thrown aside, but reverted to and relearned. The spiritual comprehension may be infinitely subtilised, but the raw material it operates upon, must remain. There may be no end of the poets who communicate to us what they see in an object with reference to their own individuality; what it was before they saw it, in reference to the aggregate human mind, will be as desirable to know as ever. (Browning 1983–2009, 4:427–8)

The world is not to be learned and thrown aside. 'This world'—this poor, mundane, material world, not as it appears when lit up by the exceptional 'individuality' of the 'subjective' poet (or medium, or clairvoyant), but as it is actually experienced by 'the aggregate human mind'—is not just a fetter binding down the exalted soul of the true visionary, as the spasmodics repeatedly asserted it to be, and as Barrett had depicted it as being in 'A Vision of Poets'. It is something *worth knowing*, and worth 'concern[ing] ourselves' with—not just as a stepping-stone to higher things, but in its own right—and to do so is not to betray one's duty as a poet, but to make a vital contribution to the ongoing development of poetry as a whole.

In a sense, Browning's 'Essay on Shelley' constituted a kind of declaration of independence from Shelley's ghost, completing a process which had begun when he begged Shelley's spirit to 'come not near' him in *Sordello*—or perhaps even earlier, when he burned 'Pauline part 2', and with it his hopes of becoming Shelley reborn. He did not question the value of Shelley's 'subjective' poetry, but he made clear that he had no intention of following its example, and spelled out why he believed his own 'objective' poetry to be just as valuable. By describing Shelley's poetry in this way, Browning was also able to finally depict Shelley himself, not as a superhuman 'sun-treader', but simply as a poet: a very great one, admittedly, but flawed and mortal none the less.

In *Pauline* Browning had written of Shelley's mental perfection, with the poem's speaker claiming that 'foul forms / Seek me, which ne'er could fasten on his mind' (Browning 1983–2009, 1:38). In his essay, however, Browning argued that Shelley's mind had not been so very perfect after all: his abilities had been constrained by the fact that his sympathies were 'developed in him to an extraordinary and even morbid degree', he had suffered from 'incurable disease', he had a life-long 'habit of somnambulism', and he had made over-heavy use of 'the laudanum-bottle' (Browning 1983–2009, 4:435, 4:439–40). He even permitted himself to critique Shelley's imagination in distinctly medicalised terms:

> He was constantly subject to "that state of mind" (I quote his own note to "Hellas") "in which ideas may be supposed to assume the force of sensation, through the confusion of thought with the objects of thought, and excess of passion animating the creations of the imagination:" in other words, he was liable to remarkable delusions and hallucinations. (Browning 1983–2009, 4:440)

A great deal of weight is borne by the phrase ‘in other words’, here, which effectively asserts that Browning’s amateur medical diagnosis of Shelley (‘remarkable delusions and hallucinations’) carries at least as much authority as Shelley’s own account of the workings of his own mind. Browning went on to argue that a whole series of bizarre events in Shelley’s life had, in fact, been ‘illusory’: they didn’t really happen, and Shelley had only thought that they did because of his ‘morbid’ sympathies and over-heated imagination. Not only did Browning no longer regard Shelley as an authority in matters of cosmic truth, he viewed Shelley’s testimony as having been rendered so doubtful by his ‘delusions and hallucinations’ that he ceased even to be an authority on his own life story.

For Browning, this unreliability was the reason why Shelley’s letters were so important: they provided corroborating documentary evidence that Shelley really had been the kind of pure and noble soul that he described himself as being in his poetry. (Whitehead has noted the extraordinary faith which biographers of the period were willing to place in letters and documents.) (Whitehead 2017, 130). Twenty years earlier Browning had been willing to take Shelley’s spiritual purity on faith, believing, like the spasmodics, that poetry such as his constituted its own proof, as only someone as noble and inspired as Shelley represented himself as having been would have been able to write it in the first place. By 1851, however, his belief in the self-validating powers of prophetic poetry had been undermined by his awareness that all kinds of charlatans, lunatics, and narcissists were capable of making equally grand claims for their own special status. Browning’s remaining loyalty to his one-time poetic idol thus came under considerable strain when, in 1852, the reassuringly high-minded collection of letters for which Browning had written the ‘Essay’ turned out to be forgeries, while in 1856 he was shown other, authentic letters which demonstrated the cruelty with which Shelley had treated his first wife, Harriet Westbrook. The best excuse he could offer for Shelley’s behaviour was that he might have been deranged at the time, presumably a victim of his ‘incurable disease’: and when, in 1885, he was offered the presidency of the Shelley Society, he refused to accept it (Keenan 1973, 119; Thomas 1982, 256; Browning 1983–2009, 4:422).

Browning’s essay makes clear the extent to which he had internalised the critiques of poetic imagination offered up by writers such as Madden, Conolly, Taylor, and Winslow over the previous quarter-century. In consequence, he had moved from regarding Shelley as so obviously inspired that everything he wrote was to be taken on trust, to regarding him as being so compromised by mental infirmities that his claims had to be regarded with great suspicion unless corroborated by external evidence. As I have argued, this deepening distrust of visionary poetry went hand in hand with Browning's scepticism regarding all other claims to special inspiration, which only hardened as Barrett Browning became ever more involved with the spiritualist movement. It is thus somewhat ironic that Moir's 1850–1851 lectures attacked Browning for his mysticism, metaphysics, and obscurity, when mere months later Browning's essay on Shelley would reveal just how much Moir's views on poetry had in common with his own (Moir 1852, 324–6). Browning differed from Moir in that he saw strongly objective and strongly subjective poetry not as being good and bad, respectively, but as necessary components of a larger whole: thus his more merciful judgement on Shelley, despite his 'delusions', 'hallucinations', and 'disease'. Yet he made clear that his own commitment was not to personal subjective vision, but to objectivity and 'the aggregate human mind': just the sort of 'general sympathy' which Moir complained was being ignored by the poets of the present day, albeit not expressed in the kind of simple, accessible poetry which Moir himself had in mind.

From Browning's perspective, the spasmodics were exactly the kind of 'tribe of successors' whom he had warned about in his essay on Shelley, 'subsisting wholly on the shadow of a reality'—the reality, in this case, being the genuine poetic achievements of Shelley and his contemporaries (Browning 1983–2009, 4:428). As Harrison notes, his 1855 poem 'Popularity' can be read as an attack on precisely this kind of poetic parasitism, expressing his irritation that the spasmodic imitators of the Romantics were more commercially successful than the Romantics themselves had ever been (Harrison 1990, 56–8). Yet, as I have mentioned, he did not wholly throw in his lot with the movement's critics, either. By 1851 he was willing to accept that Shelley's mind had been abnormal and diseased, but for him it did not follow that all Shelley's works were thus valueless, or that his influence on subsequent poetry had been entirely unhealthy, as Moir and Kingsley both maintained.

Indeed, I believe that the form of dramatic poetry which Browning perfected in *Men and Women* (1855) represented precisely an attempt to navigate between these two poles. Moir had argued that the task of the true poet was to uphold the same standards of sanity which would be recognised by a 'jury of common-sense men': divergence from this standard implied that the poet was at fault, and possibly mentally unsound. The spasmodics had asserted the reverse: that the poet's task was to articulate their own subjective insights, regardless of how odd they might look to ordinary people, trusting in their belief that the personal inspiration of the true genius transcended ordinary judgement. But Browning's dramatic monologues do neither of these things. They articulate the private, subjective perspectives of their speakers, some of whom turn out to be very odd people indeed; but rather than using his poetic authority to endorse or condemn them, Browning offered their words up to his readers to be evaluated and interpreted in whatever manner they saw fit. In Browning's terms, they constituted an objective poetry of subjectivity, which presented the words of his speakers not as authoritative truths to be accepted but as 'objects for men's outer and not inner sight', and thus side-stepped the problems of authority which did such harm to the spasmodics (Browning 1983–2009, 4:428).

The spasmodics prized the poetry of pure subjectivity above all else, but by 1853 Browning had come to feel that too much subjectivity could be an active disadvantage to a would-be poet. During his stay in Florence he had become friends with Frederick Tennyson, and in March 1853, he noted that Frederick had all the qualities that might make a poet. Frederick, he wrote, 'is to the Poet what a solution is to the chrystals which it ends in producing—the same temperament of thought & feeling, unconcentrated by the accident of the *wire*' (Kelley et al. 1984–2016, 19:18). By June, he believed that he had identified the missing '*wire*' which prevented Frederick's poetic 'temperament' from actually crystallising into poetry:

> He wants the dramatic power of his brother entirely—writes wholly *subjectively*: that purely veracious, earnest and simple mind of his cannot be other than itself for a moment. (Kelley et al. 1984–2016, 19:128–9)

The very qualities which, by spasmodic standards, should have qualified Frederick to be a great poet—his otherworldliness, his love of poetry and mysticism, and his intense subjectivity—become, in Browning's account, the reasons why he was unable to accomplish anything, either in poetry or in any other field. Describing Frederick in terms strongly reminiscent of his own Sordello, Browning wrote: 'He will stand aside all his life and see others act—yet he could not himself act, even as well, without an effort of which he is quite incapable' (Kelley et al. 1984–2016, 19:129).

The mind of Alfred Tennyson, by contrast, *did* possess the ability to be 'other than itself'. This was what Browning valued most about his poetry, and in 1843 he had expressed his frustration at the failure of Tennyson's reviewers to understand that most of Tennyson's poems, like his own, were '*Dramatic*—utterances coloured by an imaginary speaker's moods' (Kelley et al. 1984–2016, 8:38–40). Rather than being 'subjective' transcripts of his own feelings, they represented 'objective' attempts to understand what someone else's subjectivity might feel like from the inside—and it was precisely this 'dramatic power' that, in Browning's view, allowed them to succeed as poetry. As Armstrong writes, in her account of the dramatic poetry of Browning and Tennyson, '[T]he poem turns its expressive utterance around so that it becomes the opposite of itself, not only the *subject's* utterance but the *object* of analysis and critique' (Armstrong 1996, 12). It was this side of Tennyson's art which was displayed in his 1855 'monodrama', 'Maud': and it is this work, with its combination of spasmodic and anti-spasmodic features, to which I shall now turn.

'Morbid Hate and Horror': Tennyson's 'Maud' and Browning's 'Childe Roland'

Aytoun's dramatic parody *Firmillian* (1854), which gave the spasmodic movement its name, had Dobell's *Balder* as its primary target: but one of its secondary targets was Tennyson, as is made clear by the fact that Firmillian is betrothed to a woman named Mariana, and kills one his victims by throwing him off the top of St Simeon's column (Aytoun 1854, 27, 38, 109). Aytoun's decision to reference these two poems in particular suggests why he might have seen Tennyson as being a kindred spirit to the spasmodics. Like *Balder* and *Night and the Soul*, 'Mariana' and 'St Simeon Stylites' were poems which explored extreme and unbalanced mental states from the perspective of their sufferers, with 'St Simeon' in particular indulging in much the same kind of spasmodic alternation between visionary rapture and despair that characterised the poet-protagonists of Bailey, Smith, Dobell, and Bigg.

As I have discussed, a decade earlier Horne and Barrett had also linked Tennyson and Bailey, seeing both as part of a poetic movement which prioritised 'ideality' and inspired imagination over dreary realism and rational judgement. It was on these grounds that Horne had praised Tennyson for the lack of realism in his early poems about women, arguing that the very fact that it was impossible to believe in 'airy, fairy Lilian' and the rest as real human beings was proof that they possessed an admirably high degree of 'ideality' (Horne 1844, 2:16–17). Tennyson had read

Bailey's *Festus* in 1846, and admired it enormously; he also read Smith's *Life-Drama*, and started (but did not finish) Dobell's *Balder* (Rudy 2009, 105–6). Despite his evident interest in spasmodic poetry, however, his own works did not share their faith in the prophetic authority of inspired genius, and even in the 1830s he had not placed his depictions of mental anguish within contexts which glorified them as part of the necessary sufferings of truly superior minds. *Night and the Soul*, *A Life-Drama*, and 'St Simeon Stylites' are all poems in which mentally unbalanced individuals describe their experiences of acute mental distress; but the nightmare-narratives of Smith and Bigg are hedged around with narrative affirmations, both from the poet-protagonists themselves and from the other characters, which validate the authenticity of the genius of their respective protagonists and the value of the poetry which they will write once they have finally transcended their states of despair. In Tennyson's poem, however, there is no supporting cast to offer such reassurances, and no larger dramatic context to confirm Simeon's status within the world—only his desperate voice, speaking into the silence, begging for confirmation that his visions have a basis in reality. Simeon's words stand alone, as those of the spasmodics did not, giving the reader no reason to believe that his visions of suffering and salvation have any truth-value whatsoever.

Tennyson's 'Maud' was published when the spasmodic movement was at its height. It was in several respects his most spasmodic poem to date, and was regarded as such by many of its original readers (Faas 1988, 139). As Tate argues:

> The influence of spasmodism on [Maud] is also traceable in two important aspects of its psychology: its status as a dramatic poem, constructed around the mind of a specific dramatic speaker, and its sustained and hectic attention to the physiological processes that guide the workings of that mind. (Tate 2012, 116)

Like *Festus*, *Night and the Soul*, *A Life-Drama*, and *Balder*, it was dramatic in form, although it took the monologue-heavy format of these spasmodic verse-dramas to their logical conclusion by being a 'monodrama' with only a single speaker throughout. Like them, its protagonist was an emotional and mentally unstable young man, much given to flights of imagination. Like *Festus*, *Night and the Soul*, *A Life-Drama*, and the projected three-part *Balder* (and, indeed, *The Dreamer and the Worker*), it described the journey of its protagonist into spiritual turmoil and then out into some form of renewal. But the shift from dramatic poetry to 'monodrama' means that 'Maud', like 'St Simeon Stylites' and unlike the works of the spasmodics, refuses to offer external corroboration of anything that its speaker has to say.

When Tennyson wrote 'Maud', he originally intended its title to be 'Maud, or the Madness', and this secondary title points to a crucial difference between 'Maud' and the spasmodic verse-dramas which, in many respects, it outwardly resembles (Ricks 1989, 235). Like them, it depicts what Tennyson called 'the history of a morbid, poetic soul'; but unlike their protagonists, with the possible exception of Balder, the speaker of Maud *is actually insane.* As Faas notes, it was the fact that 'Maud' made clear that its speaker was mad that set it apart from *Balder*, which—despite Dobell's claims to the contrary—did nothing to undermine the apparent authority of its protagonist (Faas 1988, 12–13, 141–3). In 'Maud', the speaker's psychological abnormality is not presented as a side effect of his genius, or the result of his elevated visions being mischaracterised as madness by a world too blind to appreciate their true sublimity. Instead, it is a real (and probably hereditary) mental disorder, of the kind which, as Mighall notes, played an increasingly significant role in the Gothic fiction of the 1850s (Mighall 1999, 78–103). As Tennyson himself wrote, the speaker of 'Maud' was 'constitutionally diseased' and 'intended to have an hereditary vein of insanity' (Tennyson 1982, 2:137–8).

As in Browning's dramatic monologues—or, indeed, the works of Poe, which Tennyson read during the composition of 'Maud'—the speaker's mental state makes itself known through the very shape of his language, as in the lines in which he recalls the night of his father's death:

> Did he fling himself down? who knows? for a vast speculation had failed,
> And ever he muttered and maddened, and ever wanned with despair,
> And out he walked when the wind like a broken worldling wail'd,
> And the flying gold of the ruined woodlands drove through the air.
>
> I remember the time, for the roots of my hair were stirred
> By a shuffled step, by a dead weight trailed, by a whispered fright,
> And my pulses closed their gates with a shock on my heart as I heard
> The shrill-edged shriek of a mother divide the shuddering night. (Tennyson 1987, 2:519–20; Blair 2006, 217)

'What! am I raging alone as my father raged in his mood?' cries the speaker, realising that he has 'muttered and maddened' just as his father used to do—his mental instability apparently passed down to his son on the very night of his death, in the form of the literally spasmodic 'shock on my heart' experienced by the listening child when he hears his mother's scream at seeing his father's corpse (Tennyson 1987, 2:523). The shock of it seems never to have left him, for the verse itself is 'shuddering', its heart-beat irregular, the distribution of stresses never following the same pattern on consecutive lines.

The 'vast speculation' which led to the possible suicide of the speaker's father was, of course, a financial one; but rather than drawing a reassuring distinction between the materialistic activities of such 'broken worlding[s]' and the ennobling spiritual and philosophical 'speculations' of the speaker, as the spasmodics liked to do, 'Maud' insistently links the two. Like the spasmodics, its speaker is sceptical of what he sees as a spiritually impoverished scientific materialism; but, unlike them, he is equally wary of unbounded poetic imagination, which he sees as likely to lead to 'folly' rather than to visionary transcendence. Instead, like Browning's Paracelsus, he aspires to tread a middle path between the two:

> The man of science himself is fonder of glory, and vain,
> An eye well-practised in nature, a spirit bounded and poor;
> The passionate heart of the poet is whirl'd into folly and vice.
> I would not marvel at either, but keep a temperate brain;
> For not to desire or admire, if a man could learn it, were more
> Than to walk all day like the sultan of old in a garden of spice. (Tennyson 1987, 2:531)

'A temperate brain', however, is exactly what he does not have; for with his fears and nightmares and 'morbid hate and horror', his suicidal fantasies and his 'nerves of pain', the speaker is clearly on the edges of madness right from the beginning of the poem (Tennyson 1987, 2:523–5, 531, 536). Horne had imagined the truly imaginative 'dreamer' to be the kind of inherently superior person who could speak and act clearly and firmly even in the middle of a nightmarish shipwreck, even though 'the great masses of mankind … are not capable, either by their original nature or intellectual and moral habits, of sustaining themselves in such scenes as this' (Horne 1851, 11, 18–19). Tennyson's over-imaginative speaker, by contrast, is incapable of maintaining such composure even during a short walk in the woods.

As with St Simeon, one consequence of the speaker's obvious mental imbalance is to render doubtful everything he has to say. Like the spasmodics, he has a great deal to say about the degraded condition of his materialistic age; but his disturbed mental condition make it possible that these are just articulations of his own warped perspective rather than the inspired jeremiads of a poet-prophet, as Tennyson himself pointed out (Day 2005, 154–5). Like them, he longs for some great genius or master-spirit who will be able to force the world to reform itself, calling out in a Carlylean vein for 'one still strong man in a blatant land', but his very next words give the game away:

> And ah for a man to rise in me,
> That the man I am may cease to be! (Tennyson 1987, 2:544)

His deepest desire is not for world-mastery but *self*-mastery—a desire for which world-mastery is simply a convenient symbol, and which is constantly thwarted by the physically defective condition of his brain and heart. Rather than transcending the material world, his mind is dragged down by the malfunctioning of his body, as when he thinks he sees Maud's ghost:

> Is it gone? my pulses beat—
> What was it? a lying trick of the brain?
> Yet I thought I saw her stand,
> A shadow there at my feet,
> High over the shadowy land. (Tennyson 1987, 2:567)

The scenes in which the speaker suffers hallucinations of Maud following him everywhere form a kind of culmination to a whole sequence of similar scenes in other poems, in which men who are mad or inspired think they see visions of women before them. As I have mentioned, in Barrett's 'Lady Geraldine's Courtship', the 'mad', 'inspired' poet sees what he believes to be a hallucination of his absent beloved, only to discover that his vision has not betrayed him after all, and she is really present. Browning's 'Mesmerism' took the same scene and unsettled it: now the speaker *believes* that he is inspired and the woman is present, but the reader is given ample reason to suspect that the woman may actually be a hallucination. 'Maud' retells the same scene yet again, but now the real woman has decisively given way to a hallucinatory 'shadow', a 'lying trick of the brain'. In lines which Tennyson borrowed and adapted from his earlier poem, 'Oh! That 'Twere Possible', the speaker describes how Maud's 'phantom' constantly follows him, despite his knowledge that it must be 'a juggle of the brain'. In 'Maud', however the optimistic lines which Tennyson added to 'Oh! That 'Twere Possible' in 1837—in which the speaker attempts to regain some control over his own mind by deliberately imagining 'a phantom fair and good' who can lead him into heaven—are not reused: instead, the speaker declines further and further into madness, until he is confined in a madhouse. Fittingly, one of his fellow patients there believes himself to be 'a lord of all things': a belief which, combined with a modicum of poetic talent, might have equipped him to become a tolerable spasmodic poet.

'Maud' ends with the speaker's apparent return to sanity; but compared to the poetry of Horne, Smith, Bailey, and Bigg, 'Maud' articulates a much gloomier view of the ability of sensitive and imaginative individuals to preserve themselves from madness. For its speaker, visionary poetic imagination is not the solution to his woes, or those of the wider world; indeed, it seems to be part of the problem, as the power of the speaker's imagination causes his preoccupation with death to be projected onto the world around him. In the woods he 'hear[s] the dead at midday moan', in the 'hollow' where his father died every echo sounds like the word 'death' to him, the sound of the tide strikes him as 'the scream of a maddened beach dragged down by the wave', and the sight of the constellation Orion sinking towards the horizon strikes him as 'Orion low in his grave'. (Given that Horne's *Orion* was one of the foundational works of the spasmodic movement, the fact that the speaker perceives Orion as dead and buried may be a hint of just how useless the spasmodic ideology is to him.) According to Romantic tradition, his habit of walking 'in the quiet woodland ways' should help to heal his mind; but whereas Wordsworth's wanderings led him to 'golden daffodils', the memory of which cheered him when he lay 'In vacant or in pensive mood', the speaker of 'Maud' finds only 'the shining daffodil dead' (Tennyson 1987, 2:519, 2:528, 2:536). 'Maud', in other words, systematically undercuts the claims of the spasmodic poems which it superficially resembles. Instead of exalting the power of the imagination to redeem the material world, 'Maud' emphasises that such imaginative power is always constrained by the physical condition of the brain and heart, and that unwise indulgence in it may lead to the ruin of both.

The universe of death which the speaker of Maud perceives all around him is reminiscent of that which appears in Browning's near-contemporary poem, 'Childe Roland', which also depicts a failure or perversion of Romanticism. As Baker notes, 'Childe Roland', like 'Maud', depicts an inverted version of Wordsworthian vision, in which mental disturbance triumphs over nature rather than the other way around (Baker 2004, 106). Both poems have classically 'Romantic' speakers—a sensitive lover crushed by the world's heartlessness and an adventurous knight-errant on a quest, respectively—but rather than showing their idealism as allowing them, like Horne's 'dreamer', to see glory and beauty in a world which strikes others as mundane, they are depicted as being dragged down by their malfunctioning imaginations into unbearable worlds of fear and death. If they had, like Browning's 'objective poet', been able to see the world as it appeared to 'the common eye and apprehension of [their] fellow men', then they might have been able to spare themselves a great deal of mental suffering (Browning 1983–2009, 4:424). But, instead, like the poet-heroes of the spasmodics, they cling to their singular visions of life—even at the cost of their own sanity.

As a result, while both poems have some traits in common with those of the spasmodics, their overall tendency is very different. The visions of universal misery which appear in 'Maud' and 'Childe Roland' resemble those of Alexis in Bigg's *Night and the Soul*, but stripped of all the consolations that Bigg offers to his readers: the assurance that Alexis is a great man and not a mere lunatic, that he will eventually emerge from his despair with his imagination revitalised and become a true poet, and that to be a true poet is a very great thing indeed. They signal, much more clearly than Dobell's *Balder*, that interpretative authority should rest not with their delusional speakers, but with the readers who evaluate their words, and in this sense they are closer in spirit to the critics of spasmodism than to the spasmodics themselves. It was for this reason that Tennyson welcomed the kind of medically informed criticism which demolished the reputation of the spasmodics, and was delighted when doctors wrote to him diagnosing exactly what had gone wrong inside the brain of the speaker of 'Maud' (Tennyson 1982, 2:131; Faas 1988, 13, 26, 32–3; Colley 1983, 49; Platizky 1989, 50). The poem was even reviewed in *The Asylum Journal of Mental Science* by John Bucknill, who praised Tennyson's 'profound knowledge of mental pathology' and congratulated him on keeping the poem 'true to psychological probabilities' (Bucknill 1855, 96, 104). For Tennyson, as for Browning, this kind of objective evaluation of a speaker's subjectivity was exactly the way in which dramatic poetry was meant to be read: and the poet, far from being the doctor's rival, was more properly his accomplice in the analysis of abnormal forms of human subjectivity. As Faas argues, 'Tennyson and Browning used the dramatic monologue to display abnormalities the way an alienist might observe and analyse a patient in his madhouse cell' (Faas 1988, 17).

Conclusion: 'A Disease of the Intellect'

By 1855—the year which saw the publication of Browning's *Men and Women* and Tennyson's *Maud and Other Poems*—the institutional success of medicalised approaches to the study and treatment of the human mind in Britain was largely complete. The Association of Medical Officers of Asylums was founded in 1841, and started to publish the *Asylum Journal* (later the *Journal of Mental Science*) in 1853; in 1865 it would change its name to 'the Medico-Psychological Association', the ancestor of the modern Royal College of Psychiatrists (Jones 1972, 155). Lunacy legislation in Britain had been enormously expanded and reformed since the Madhouses Act of 1828—largely under the guidance of Lord Shaftesbury, who was so devoted to the issue that he was rumoured to be a lunatic himself—and the M'Naghten rules of 1843 and the 1845 Asylums Act had hugely increased the authority of the medical profession over the identification and treatment of the insane (Skultans 1979, 98). The year 1851 saw the opening of the enormous new Colney Hatch asylum outside London—the largest and most modern institution of its kind in Europe, with space for 1250 inmates, and built at the eye-watering cost of £300,000. (In 1858, its capacity was expanded even further, to 2000 [Showalter 1987, 23, 101].) Conolly, who in the 1830s had been in the vanguard of the movement for non-restraint and asylum reform, had by the 1850s concluded that Britain's new asylums were just the kind of institutions he had once dreamed of. Towards the end of his life he became infamous for his willingness to certify anyone even mildly odd or unruly, on the assumption that the new asylums were such splendid institutions that almost anyone might benefit from a stay inside one (Showalter 1987, 47–8; Scull et al. 1996, 75–6, 80).

The increasing cultural authority of these discourses impacted, in turn, upon the way in which 'visionary' poetic inspiration was understood during this period. As Shelley was generally regarded as a visionary poet *par excellence*, the changing ways in which his life and works were understood can be taken as a marker of these shifting views: and, as I have discussed, by the 1850s even writers such as Moir, Browning, and Kingsley, who were otherwise sympathetic to much of what would now be called 'Romantic' poetry, were describing Shelley and his visions in medicalised and pathological terms. Given all this, it would be tempting to conclude that 'Maud' and 'Childe Roland' mark the point at which the new discourses of psychiatry triumphed over faith in 'inspired' Shelleyan poetry, even among those writers who had once been most sympathetic to it. The spasmodics presented the authority of inspired poets as transcending that of doctors, and were mocked (and pathologised) for it. Browning and Tennyson presented men of imagination as though they were psychological case studies, and went on to become two of the most prominent and highly regarded poets of the age. Further evidence might be adduced from the changes which Tennyson made to *The Princess* in 1851, adding new and apparently unnecessary passages to the poem in which its speaker confessed that he suffered from 'weird seizures' during which he felt himself to be 'the shadow of a dream', as though undermining the narrative authority of his more 'Romantic' speakers by implying them to be physically and mentally unsound had by this point become something that Tennyson did almost reflexively (Faas 1988, 76).

The reviews which greeted these works, however, indicate that the truth is less straightforward. Far from being celebrated as antidotes to the excesses of the spasmodics, 'Maud' and *Men and Women* were widely attacked as being, themselves, examples of spasmodism (Harrison 1990, 57–9). Some readers of 'Maud' congratulated Tennyson on his brilliantly accurate depiction of madness, but others condemned him for writing yet another spasmodic poem in celebration of dangerous mental abnormality, and Tennyson complained for decades about 'the somewhat perverse allegations that [he] himself simply spoke his own thoughts through the madman-hero of "Maud"' (Tennyson 1897, 2:498). Browning's critics accused him of madness: *The Critic* suggested that *Men and Women* showed Browning to be suffering from 'a disease of the intellect', while *The Morning Post* accused him of writing 'under the influence of *spasm*'. *The Saturday Review* described *Men and Women* as being 'another book of madness and mysticism', comparing it to 'Alexander Smith, or Alfred Tennyson's *Maud*, or the *Mystic* of Bailey', and remarking that 'It is really high time that this sort of thing should, if possible, be stopped' (Kelley et al. 1984–2016, 22:314, 22:329, 22:302). *The Press* scornfully awarded Browning 'foremost place in the fragmentary or spasmodic school', before linking his works together with Tennyson's 'Maud':

> To be sure, the defence which has been set up for "Maud," may be set up for "Men and Women." If there be men of hysterical temperament, why should we not have a hysterical poem; and if there be men of affected intellect, why not have affected verse? For no other reason than this, that the aim of poetry should be to elevate or refine the mind, never to dwarf or debase it. (Kelley et al. 1984–2016, 22:296–7)

For this reviewer, to write a poem such as 'Maud' or 'Childe Roland', in which a 'hysterical' speaker was allowed to speak for themselves rather than having their words safely contained within a more 'elevate[d] or refine[d]' discourse, was to misunderstand or betray the educational purpose of poetry. Such poems aligned the poet, not with the voice of reason and sanity—the voice of the doctor, one might say—but with that of the lunatic. Thus the reviewer's sneer that 'the annals of Bedlam could supply volumes of matter of very similar kind—proofs of the poetic inspiration of its inmates' (Kelley et al. 1984–2016, 22:297).

If Browning and Tennyson had wished to avoid such criticism, they could easily have done so by writing poems that explained and demonstrated exactly what was wrong with their mentally unbalanced protagonists. But neither poem does this: and while 'Maud' and 'Childe Roland' do not attempt to insist upon the transcendent authority of the subjective visions of their respective speakers, neither do they reductively present those visions as mere symptoms of mental pathology. The criticisms levelled by Moir at Browning, and by Aytoun at Tennyson, demonstrate that neither man was regarded as entirely free from spasmodism, and the refusal of these poems to come down clearly on one side of the debate or the other indicate part of the reason why. They 'introduce the possibility of interrogation and critique', as Armstrong notes, but they do not attempt to predetermine what the result of such critiques should be (Armstrong 1996, 12). Instead, their speakers are allowed to speak without any overbearing interventions from their authors to indicate to the reader that *this* is to be read as genuine visionary inspiration while *that* is to be understood as pathological insanity.

The problem which most vexed the reviewers of these poems was what they called 'the obscure': the way in which their meanings had to be puzzled out by the reader, rather than being straightforwardly presented to them by the poet (Kelley et al. 1984–2016, 22:296). Yet it was precisely this reallocation of interpretative authority which offered Tennyson and Browning a way out of the impasse between the spasmodics and their critics. If the hero of a poem such as *Night and the Soul* or *A Life-Drama* is read as being mad rather than inspired, then the poem as a whole will misfire, as every part of it which is meant to be sublime will seem absurd or bathetic, instead. (As I have discussed, this was the basis of Aytoun, Arnold, Moir, and Kingsley's respective critiques of spasmodism.) *Balder* failed precisely because many of its readers felt that the poem expected them to view its protagonist as inspired, even though they actually regarded him as insane. But it matters very little whether the speakers of 'Maud' or 'Childe Roland' are read as being mad, or inspired, or both simultaneously, because the purpose of these poems is to communicate the nature of certain subjective experiences to their readers, rather than to persuade them that the conclusions reached by their speakers are valid and true. These poems did not lay claim to the kind of visionary authority that Tennyson and Browning had longed for as adolescents, and which Bailey and his imitators still hungered after in the 1850s. But they did claim a new and different kind of poetic authority: the authority to accurately (Browning might have said 'objectively') evoke and communicate unusual modes of consciousness, even as the task of determining whether those modes were to be classed as 'mad' or 'sane' was increasingly relinquished to doctors rather than poets.

This shift, I believe, was one important result of the generation-long struggle to define the relationship between genius and the mental infirmities which so often seemed to accompany it. The medicalisation and pathologisation of poetic genius made it increasingly difficult for writers to credibly claim the kind of prophetic authority which their predecessors had aspired to: as I have suggested, spasmodic poetry can be understood as something of a rearguard action in this respect, while the popularity of mesmerism among post-Romantic poets may have stemmed from its apparent ability to provide objective proof that they really *did* possess mental and spiritual powers beyond those of ordinary people. But as mesmerism and spasmodism alike fell out of favour in the later 1850s, outgunned by an increasingly authoritative 'medico-psychological' establishment, the dramatic monologue form which Browning and Tennyson had jointly developed pointed another way forwards, by breaking the assumed connection between subjectivity and authenticity that the great Romantic lyrics had helped to usher in a generation before.

From this perspective, 'Maud' and 'Childe Roland' can be seen as culminations of the decades-long process by which Browning and Tennyson had sought to find places for themselves as poets within a culture that was growing ever more wary of the supposed infirmities of genius. Through their objective presentation of subjective mental states, depicted as explorations of possible configurations of human psychology rather than as authentic reflections of the poet's own state of mind, such poems were able to retain a classically 'Romantic' fascination with unusual and visionary states of consciousness while also withstanding critiques informed by contemporary medical psychology without self-destructing. For readers and critics who still reflexively assumed that the voice of a poem's speaker must also be the voice of the poet, such poems could only ever serve as so many bewildering proofs of 'a disease of the intellect'. But as their essentially dramatic character came to be better understood, they went on to provide a template for the similarly 'psychological' poetry of writers such as Morris, Swinburne, and the Rossettis, long after the grand claims of spasmodism had been allowed to fall into silence.

References

Primary Sources

Aytoun, William Edmondstoune. 1854. *Firmilian: A "Spasmodic" Tragedy.* New York: Redfield.

Bailey, Philip James. 1845. *Festus: A Poem.* Boston: Benjamin Mussey.

Barrett Browning, Elizabeth, and Robert Browning. 1990. In *The Courtship Correspondence 1845–6*, ed. Daniel Karlin. Oxford: Oxford University Press.

Bigg, John Stanyan. 1854. *Night and the Soul.* London: Groombridge and Sons.

Brisbane, Thomas. 1869. *The Early Years of Alexander Smith, Poet and Essayist.* London: Hodder and Stoughton.

Browning, Robert. 1983–2009. *Poetical Works*, eds. Ian Jack, Rowena Fowler, Margaret Smith, Robert Inglesfield, et al., 15 vols. Oxford: Clarendon.

Bucknill, John Charles. 1855. Review of *Maud and Other Poems. The Asylum Journal of Mental Science* 2 (15): 95–104.

Carlyle, Thomas. 1866. *On Heroes, Hero-Worship, and the Heroic in History.* New York: John Wiley and Son.

———. 1975. *Sartor Resartus.* London: J.M. Dent and Sons.

Dobell, Sydney. 1854. *Balder.* 2nd ed. London: Smith, Elder, and Co.

Eliot, George. 1999. In *The Lifted Veil and Brother Jacob*, ed. Helen Small. Oxford: Oxford University Press.

Gilfillan, George. 1851. *The Bards of the Bible.* New York: Harper and Brothers.

Horne, Richard Henry, ed. 1844. *A New Spirit of the Age.* 2 vols. London: Smith, Elder, and Co.

———. 1851. *The Dreamer and the Worker.* 2 vols. London: Henry Colburn.

Kelley, Phillip, et al., eds. 1984–2016. *The Brownings' Correspondence.* 23 vols. Winfield: Wedgestone.

Kingsley, Charles. 1853. Thoughts About Shelley and Byron. In *Fraser's Magazine for Town and Country*, vol. 48, 568–576. London: John W. Parker and Son.

———. 1906. *Two Years Ago.* London: Macmillan and Co.

Moir, D.M. 1852. *Sketches of the Poetical Literature of the Past Half Century, in Six Lectures.* 2nd ed. Edinburgh: William Blackwood and Sons.

Searle, January. 1852. *Memoirs of William Wordsworth*. London: Partridge and Oakey.

Shelley, Percy Bysshe. 1977. *Poetry and Prose*, ed. Donald Reiman and Sharon Powers. New York: Norton.

Smith, Alexander. 1859. *A Life-Drama and Other Poems*. Boston: Ticknor and Fields.

———. 1863. *Dreamthorp*. London: Strahan and Co.

Tennyson, Alfred. 1982. *Letters*, ed. Cecil Lang and Edgar Shannon Jr., 3 vols. Oxford: Clarendon.

———. 1987. *The Poems of Tennyson*, ed. Christopher Ricks, 3 vols, 2nd ed. Harlow: Longman.

Secondary Sources

Armstrong, Isobel. 1996. *Victorian Poetry*. London: Routledge.

Baker, John. 2004. *Browning and Wordsworth*. Madison: Fairleigh Dickinson University Press.

Blair, Kirstie. 2004. Spasmodic Affections: Poetry, Pathology, and the Spasmodic Hero. *Victorian Poetry* 42 (4): 473–490.

———. 2006. *Victorian Poetry and the Culture of the Heart*. Oxford: Clarendon.

Boos, Florence. 2004. "Spasm" and Class: W. E. Aytoun, George Gilfillan, Sydney Dobell, and Alexander Smith. *Victorian Poetry* 42 (4): 553–584.

Colley, Ann. 1983. *Tennyson and Madness*. Athens, GA: University of Georgia Press.

Coyer, Megan. 2016. *Literature and Medicine in the Nineteenth-Century Periodical Press: Blackwood's Edinburgh Magazine, 1817–1858*. Edinburgh: Edinburgh University Press.

Day, Aidan. 2005. *Tennyson's Scepticism*. Houndmills: Palgrave Macmillan.

Faas, Ekbert. 1988. *Retreat into the Mind: Victorian Poetry and the Rise of Psychiatry*. Princeton: Princeton University Press.

Gill, Stephen. 1998. *Wordsworth and the Victorians*. Oxford: Clarendon.

Haigwood, Laura. 1986. Gender-to-Gender Anxiety and Influence in Robert Browning's *Men and Women*. *Browning Institute Studies* 14: 97–118.

Harrison, Anthony. 1990. *Victorian Poets and Romantic Poems*. Charlottesville: University Press of Virginia.

Jones, Kathleen. 1972. *A History of the Mental Health Services*. London: Routledge and Kegan Paul.

Keenan, Richard. 1973. Browning and Shelley. *Browning Institute Studies* 1: 119–145.

Matthew, H.C.G., and Brian Harrison, eds. 2004. *Oxford Dictionary of National Biography*. 60 vols. Oxford: Oxford University Press.

Matthews, G.M., ed. 1971. *Keats: The Critical Heritage*. London: Routledge and Kegan Paul.
Mighall, Robert. 1999. *A Geography of Victorian Gothic Fiction: Mapping History's Nightmares*. Oxford: Oxford University Press.
Mitchell, Leslie. 2003. *Bulwer Lytton: The Rise and Fall of a Victorian Man of Letters*. London: Hambledon and London.
Morton, Heather. 2008. The 'Spasmodic' Hoaxes of W.E. Aytoun and A.C. Swinburne. *SEL* 48 (4): 849–860.
Platizky, Roger. 1989. *A Blueprint of His Dissent: Madness and Method in Tennyson's Poetry*. Lewisburg: Bucknell University Press.
Ricks, Christopher. 1989. *Tennyson*. 2nd ed. Berkeley: University of California Press.
Rudy, Jason. 2009. *Electric Meters: Victorian Physiological Poetics*. Athens: Ohio University Press.
Rutherford, Andrew, ed. 1970. *Lord Byron: The Critical Heritage*. London: Routledge and Kegan Paul.
Ryan, Robert. 1997. *The Romantic Reformation: Religious Politics in English Literature, 1789–1824*. Cambridge: Cambridge University Press.
Scull, Andrew, Charlotte Mackenzie, and Nicholas Hervey. 1996. *Masters of Bedlam: The Transformation of the Mad-Doctoring Trade*. Princeton: Princeton University Press.
Showalter, Elaine. 1987. *The Female Malady: Women, Madness, and English Culture, 1830–1980*. London: Virago.
Skultans, Vieda. 1979. *English Madness: Ideas on Insanity, 1580–1890*. London: Routledge and Kegan Paul.
Tate, Gregory. 2012. *The Poet's Mind: The Psychology of Victorian Poetry 1830–1870*. Oxford: Oxford University Press.
Tennyson, Hallam. 1897. *Alfred Lord Tennyson: A Memoir*. 2 vols. London: Macmillan.
Thomas, Donald. 1982. *Robert Browning: A Life Within Life*. New York: Viking.
Tucker, Herbert. 2004. Glandular Omnism and Beyond: The Victorian Spasmodic Epic. *Victorian Poetry* 42 (4): 429–450.
Whitehead, James. 2017. *Madness and the Romantic Poet: A Critical History*. Oxford: Oxford University Press.

CHAPTER 7

Epilogue: 'It is strange.'

In *The Poet's Mind*, Tate writes:

> The impact of psychological science can be traced through the growing willingness of critics to perceive affinities between the work of poets and the analytical practices of psychologists. A minority of commentators had argued in support of self-analytic poetry in the first half of the century, but it was only in the 1850s and 1860s that significant numbers began to applaud the use of poetry as a vehicle for psychological introspection and to argue that psychological theory had made a positive contribution to poetic practice. (Tate 2012, 16–17)

Looking back from the vantage point of 1869, H.B. Forman argued that the rise of what he called 'a definitive and invaluable science of psychology' in Britain had exerted a powerful influence over the development of British literature. It had, he wrote, led to the appearance of a new 'Psychological School of Poetry', typified by the use of the dramatic monologue as a kind of psychological case study (Faas 1988, 3–4). Over the course of this book, I have endeavoured to trace the emergence of this 'Psychological School'—not simply as the result of a straightforward application of the 'science of psychology' to poetry, as Forman implied, but also as a response to it, as poets such as Browning and Tennyson sought for ways to reconcile the claims of this new science with the faith in visionary inspiration that they had inherited from the Romantics. As I

© The Author(s) 2019

J. Crawford, *Inspiration and Insanity in British Poetry*, Palgrave Studies in Literature, Science and Medicine, https://doi.org/10.1007/978-3-030-21671-9_7

have argued, the mature dramatic monologue provided a way of presenting unusual mental states in poetry while leaving the task of interpreting those states to the reader, thus evading the problems inherent in spasmodic poetry, which could only function effectively if the inspired status of both poet and protagonist was taken on trust.

Forman saw this new school of poetry as a definite advance: and just as the asylum reformers hoped that the new sciences of the mind would permit madness to be identified and treated more effectively than had previously been possible, so Forman suggested that these same sciences would allow poets to describe human psychology more powerfully and precisely than ever before. As I have discussed, however, far from bringing enhanced certainty and confidence, the chief hallmark of this new school of poetry seems to have been a pervasive and corrosive sense of doubt. By insisting that poetic genius was, itself, a form of psychological abnormality akin to madness, Forman's 'science of psychology' left deeply unclear the extent to which poetic imagination and inspiration could be trusted by their possessors. Indeed, as I have argued, it was precisely this uncertainty which formed the basis for some of the most influential British poetry of the period.

Having reached such a point, what place remained for the genuine visionary? Browning worried away at this question in several of the poems in *Men and Women*. In 'Cleon', for example, he imagined how the doctrines of Christianity would have sounded to a sophisticated intellectual of the Classical world:

> Thou canst not think a mere barbarian Jew,
> As Paulus proves to be, one circumcised,
> Hath access to a secret shut from us?
> Thou wrongest our philosophy, O king,
> In stooping to inquire of such an one,
> As if his answer could impose at all!
> He writeth, doth he? well, and he may write.
> Oh, the Jew findeth scholars! certain slaves
> Who touched on this same isle, preached him and Christ;
> And (as I gathered from a bystander)
> Their doctrine could be held by no sane man. (Browning 1983–2009, 5:429)

Fully convinced of his intellectual superiority, Cleon dismisses St Paul's teachings as mere lunacy, unworthy of serious consideration. Significantly, his reaction to hearing about the preaching of the Christians—'Their doctrine could be held by no sane man'—is the same judgement which Browning's own contemporaries made upon a whole range of would-be visionaries in their own day, from self-proclaimed prophets to mesmerists and spiritualists to aspiring Romantic poets. Cleon makes this judgement confidently because he believes that his 'philosophy', like the 'definitive' psychological science celebrated by Forman, is so advanced that he cannot believe 'a mere barbarian Jew' could possibly '[have] access to a secret shut from us'. Instead, like the sceptical writers of Browning's day, from Hibbert in the 1820s to Madden and Moir in the 1850s—or, indeed, like the journalists who commented on John Tom's 1837 rebellion—Cleon considers it much more likely that Paul and his followers are simply insane, especially given their low social status. Yet for Browning and his audience, most of whom were observant Christians, Cleon's judgement represents a tragic error, as his over-confidence in his ability to accurately distinguish between vision and madness costs him what may well be his only chance of salvation.

'Cleon', then, illustrates the danger that such scepticism, if taken too far, might actually become a form of credulity, in which the sufficiency and superiority of medicalised explanations for unusual human behaviour is simply taken on faith. (In this context, it is noteworthy that Cleon, despite boasting of his 'philosophy', is willing to write the Christians off as lunatics based only on the second-hand reports of their preaching 'gathered from a bystander'.) In this sense, it forms a companion piece to another poem from Browning's 1855 collection, 'An Epistle Containing the Strange Medical Experience of Karshish'. The 'Epistle' describes how a first-century Arab physician, investigating the aftermath of the raising of Lazarus, concludes that Christ was merely a skilful healer, lynched by his superstitious countrymen who mistook his cures for magic. Karshish's confidence in his own medical training, like Cleon's in his sceptical philosophy, lead him to dismiss the possibility that a real miracle may have taken place: for him, Lazarus is 'stark mad', the victim of 'a case of mania, subinduced / By epilepsy'. Lazarus tells Karshish the story of Christ's incarnation, which he finds both moving and deeply troubling, but his belief in medicalised explanations will not allow him to accept that it might be true. In the end, he can only conclude helplessly that 'The madman saith He said so: it is strange' (Browning 1983–2009, 5:93–102).

Karshish's reasoning, here, is very much in line with that of the medical writers of Browning's own day. As I have discussed, their sceptical, pathologising perspective was one with which Browning had tended to ally himself over the previous two decades, as he positioned himself in opposition to the mesmeric and spiritualistic enthusiasms of the period. However, this poem, like Browning's *Paracelsus* 20 years earlier, draws attention to the limitations of such a reductive position, as Browning's scepticism wrapped around to critique the limitations of scientific rationalism just as it had critiqued those of visionary inspiration a generation before. Unlike Cleon, Karshish does not fail to recognise the truth out of simple intellectual arrogance: he employs due diligence in his investigation of the raising of Lazarus, and it is hardly his fault that he is not persuaded of Christ's divinity. (The spasmodics had clung to the idea that if a literary work was sufficiently inspired, then its prophetic authority would be obvious: but in these poems, even the words of the risen Lazarus and the Apostle Paul fail to persuade their listeners.) Karshish's sceptical conclusions may be incorrect, but they seem perfectly reasonable given the information available to him. The only hint that he may have been mistaken is the irreducible residue of *strangeness* that remains.

'It is strange.' Karshish is not wrong, here: it *is* strange, and it *remains* strange even after he has assigned it a supposedly rational explanation. Some of the same strangeness clings to many of the phenomena which the medical psychology of the post-Romantic period was called upon to explain: Irvingite ecstatics speaking in tongues, spiritualists talking to the dead, mesmerists healing diseases by staring at people, John Tom's followers charging suicidally towards the queen's soldiers, spasmodic poets solemnly declaring that poetry could save the world and make them the kings of the earth. Armstrong notes that 'madness opens up as a condition imputed to any state which cannot be culturally assimilated', but Browning's poems suggest that the *strangeness* of such states exceeds the rational explanations that are meant to account for and dispel it (Armstrong 1996, 310). It is precisely this strangeness, this residual sense of the weird or the uncanny which clings to such events regardless of whether or not we have 'explained' them, that so much of Browning's and Tennyson's poetry evokes. Rational explanations can be assigned to Paul and Christ and Lazarus, just as they can be to Childe Roland, and the speaker of 'Maud', and St Simeon Stylites, and Porphyria's lover: but their strangeness persists none the less, preserved and articulated through the poetic qualities of the language in which their stories are told.

Over the course of this book I have outlined the challenge which the development of nineteenth-century medical psychology posed to the

British poets of the post-Romantic era, due to its insistence that poetic genius and visionary inspiration were best understood in medicalised and pathological terms. As I have discussed, both mesmerism and spasmodism appealed to those poets who attempted to resist this encroachment upon what they saw as their proper authority; but the 'psychological school' of poetry identified by Forman, which Tennyson and Browning did so much to create, demonstrated what ultimately proved to be a more viable way for poetry and medical psychology to coexist. Over the period from 1830 to 1855, both writers increasingly abandoned their old prophetic aspirations: Gill notes that Tennyson, unlike Wordsworth, was 'acutely aware of his inadequacy to deal with the whole compass of the universe', while Martens describes Browning as having been 'self-conscious enough to realize that even the seemingly so authoritative figure of the poet could not lay claim to an absolute perspective' (Gill 1998, 203; Martens 2011, 20). As a result, it was less the absolute truth of things than their *strangeness* that both Tennyson and Browning ultimately came to articulate in their mid-century poetry. The objective truth about unusual states of consciousness could be left to doctors and scientists, whose growing institutional authority increasingly allowed them to pull rank over mere poets, but the articulation of the subjective strangeness of such mental states was something at which the formal resources of poetry still allowed it to excel.

Unlike the poems of the spasmodics, or the grand claims of the mesmerists and spiritualists, the 'psychological' works that Browning and Tennyson produced made no assertions about the objective validity of the states of mind that they described—and, as a result, they were very much harder to dismiss as mere side effects of the infirmities of genius. The experience of reading a poem such as 'Maud' or 'Childe Roland' is quite unlike the experience of reading case studies from contemporary works on madness, even when the symptoms described are identical. The poems communicate something of what it might be like to inhabit such minds *from the inside*, and the unsettling quality of this experience remains regardless of whether or not their speakers are ultimately classed as being inspired or insane. And if the reality of inspiration is no longer insisted upon, neither is it ruled out. Some hope remains, however chastened, that it may still exist, waiting and possible, somewhere beyond the borders of the printed page.

Sometimes the 'barbarian Jew' might really be a prophet.

What looks like a 'case of mania' may actually be a genuine resurrection.

Sometimes infirmity and genius are not so easily disentangled.

'The madman saith He said so: it is strange.'

References

Primary Sources

Browning, Robert. 1983–2009. *Poetical Works*, eds. Ian Jack, Rowena Fowler, Margaret Smith, Robert Inglesfield, et al., 15 vols. Oxford: Clarendon.

Secondary Sources

Armstrong, Isobel. 1996. *Victorian Poetry*. London: Routledge.

Faas, Ekbert. 1988. *Retreat into the Mind: Victorian Poetry and the Rise of Psychiatry*. Princeton: Princeton University Press.

Gill, Stephen. 1998. *Wordsworth and the Victorians*. Oxford: Clarendon.

Martens, Britta. 2011. *Browning, Victorian Poetics, and the Romantic Legacy: Challenging the Personal Voice*. Farnham: Ashgate.

Tate, Gregory. 2012. *The Poet's Mind: The Psychology of Victorian Poetry 1830–1870*. Oxford: Oxford University Press.

Index

© The Author(s) 2019

J. Crawford, *Inspiration and Insanity in British Poetry*,
Palgrave Studies in Literature, Science and Medicine,
https://doi.org/10.1007/978-3-030-21671-9

FSC
www.fsc.org
MIX
Papier | Fördert
gute Waldnutzung
FSC® C083411

Kolibri 360 GmbH
Ferdinand-Jühlke-Straße 7
99095 Erfurt, Deutschland
produktsicherheit@kolibri360.de